CW01370077

JAGUAR
Mark VII to 420G
The Complete Companion

JAGUAR
Mark VII to 420G
The Complete Companion
NIGEL THORLEY

BAY VIEW BOOKS

Published 1994 by Bay View Books Ltd
Glen Firs, Glentorr Road
Bideford, Devon EX39 2LS

© Copyright 1994 by Bay View Books Ltd
ISBN 1 870979 41 9
Typesetting and layout by Chris Fayers
Printed in Hong Kong by Paramount Printing Group

Contents

Before the twin cam: the Mark V Jaguar
6

Introducing the Mark VII
16

Development of the Mark VII
34

Mark VIII More luxury, more power
53

"And now the Mark IX…"
62

Mark VII-IX Specials and One-offs
77

A new generation: The Mark X
88

Improving the Mark X
108

The 4.2-litre Mark X
117

Last of the line: The 420G
124

Mark X Specials and One-offs
137

Daimler DS420, the Mark X in new clothes
141

Purchase, Maintenance and Restoration
148

Big Jaguars in Competition
163

Specifications, Performance and Production Details
170

Useful Addresses
174

Before the twin cam:
the Mark V Jaguar

Sir William Lyons standing outside the Jaguar Factory in Browns Lane, Coventry, alongside Mark X saloons awaiting distribution.

Although the subject of this book is the flagship saloons of the Jaguar marque from the Mark VII to the 420G, it is necessary to turn the clock back to the earlier days of SS and Jaguar to reflect on the development of the Jaguar saloon and in particular one model: the Mark V from the immediate post-war period.

Disregarding at this stage the original Swallow and SS cars, effectively the first saloons produced by the SS company were the 1½- and 2½- litre Standard-engined models of 1935. Utilising bought-in engines and chassis with traditionally built steel bodies around ash frames, these cars were well designed, rakish in appearance and decidedly value for money at £295 and £385 respectively.

By September 1937 Jaguar had adopted a new bodywork manufacturing method involving all-steel construction without wooden framing. In this way costs were cut and production increased, allowing the SS Jaguars to continue in production up to the outbreak of the Second World War, with the addition of a 3½-litre model. After the war production of the same three models restarted with minor modifications, and continued until the end of 1948, by which time a total of just over 26,000 had been produced.

The SS Jaguar saloons of this period had proved to be a success for the company, yielding a good profit through reasonable production levels and interchangeability of parts throughout the range. Jaguar was already recognised as a manufacturer of quality saloons for the middle classes alongside and the equal of such established greats as Rover, Riley and Humber.

After the war most manufacturers were looking at producing new models, which would revitalise the motor industry and earn important money for Britain through exports. In the short term, however, they recommenced by producing pre-war designs, and here William Lyons was no exception. Due to the high profile of the SS insignia during the war Lyons dropped this desig-

The mid-1930s SS saloon, and the very successful SS100 sports car.

nation and the company was renamed Jaguar Cars Limited. Thus Jaguar was born as a marque in its own right.

By the end of 1947 the company had a brand new engine, chassis and front suspension system, but was still developing an entirely new bodyshell to accommodate them. The new bodyshell would be all enveloping, necessitating the use of large presses to produce the complex panels needed – production methods that Jaguar could not provide in-house. The final development of the new body would take a long time to perfect.

In the interim period Jaguar decided to launch a reworked version of the pre-war cars (Mark IVs as they were later to be known), allowing the use of the factory's own body assembly facility but at the same time utilising the new chassis and front suspension. The company felt that it would be wrong at this stage to fit the brand new six-cylinder

Clay design for a new Jaguar post-war saloon that never progressed further.

engine, not only because it would perhaps steal the thunder from the new saloon when launched later, but also because production of the new unit could not be geared up quickly enough to make sufficient numbers for a saloon as well as for the new XK120 sports car, which would also use this engine. Thus the Mark V saloon was born – Mark V being the designation for no apparent reason other than the fact that it was the fifth prototype developed!

Released to the public at the Motor Show in October 1948, the Mark V may have looked very much like the older models but under the skin it was a very different motor car indeed. *The Autocar's* introductory report quoted the Mark V as having "Individual Character Enhanced in New Styling".

The chassis of the new car was an excellently designed structure, apparently one of the strongest and most rigid of its day, made up of straight box-section side-members reaching a maximum section size of 6½in deep by 3½in wide below the bulkhead. Cross-bracing was featured in the centre section for extra strength. Rearward the chassis swept up 18in over the rear axle before tapering and dropping away to the rear spring hangers. This gave good support to the rear of the body and at the same time allowed extra suspension movement. At the front a heavy box-section crossmember gave support to the new front suspension.

The other significant change was the adoption for the first time on any Jaguar of independent front suspension (a rarity on many other conventional saloons of the period). Developed over some years going back to 1938, the Jaguar unit was made up of forged upper and lower wishbones with longitudinal torsion bars, ball-jointed stub axles, and Newton telescopic shock absorbers.

The lower wishbones took the form of an I-section beam carrying the load from the front wheels to the torsion bars attached below the chassis. The upper wishbones were anchored on a bracket above the chassis. Ball joints carried the swivel assembly on the stub axle. The 52in-long torsion bars were made from silicon manganese spring steel with splined ends, the forward end fitting to the lower wishbone via Metalastic bushes. The rear end was fixed at an adjustable anchorage close to the middle of the chassis side-members. The system was so exceptionally well designed that it would continue in use with minor modifications throughout the production of the Mark VII to IX saloons and all XK sports cars. Burman recirculating ball steering continued in use although with a revised 3¼ turns lock to lock.

The rear suspension was conventional, with softer, 6in longer springs allowing for greater wheel movement than on the previous models. Girling lever arm shock absorbers were employed. The early Mark Vs had an ENV rear axle but later models used a more up-to-date Salisbury unit.

A hydraulic braking system, developed by Girling, was used for the first time on any SS/Jaguar, with a chassis-mounted single master cylinder operating a two leading shoe system. The

The swooping lines of the Mark V echoed those of the model it succeeded.

The Mark V's clubland interior would make any Jaguar owner feel at home.

The slab-sided appearance of the Mark VII was significantly improved by the stylised swage line.

Full rear spats added to the air of sophisticated elegance.

Original Mark VII frontal aspect. Note the inset fog lamps and period wing mirrors.

The unusual bonnet badge on Mark VII and VIIM depicting a head swept back into the centre chrome trim. It never appeared on any other Jaguar model.

The original small rear lights with separate reflectors as fitted to the very earliest Mark VIIs, later modified to the chrome plinth type incorporating the reflector.

11

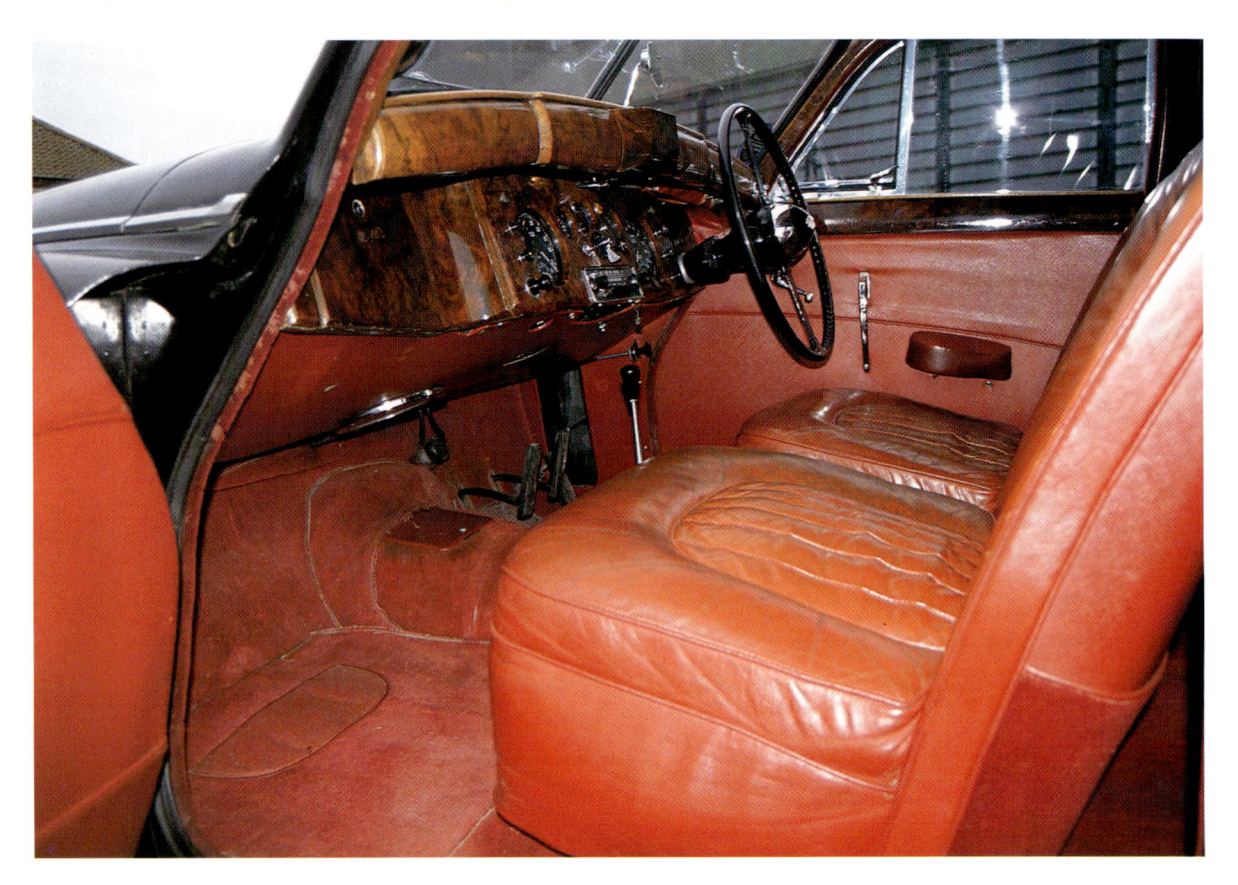

Front compartment of an early Mark VII shows the substantial leather seating and handsome dashboard. This manual transmission model also sports its original HMV valve radio, with large-diameter speaker below the centre dash section.

Luxurious rear compartment of an early Mark VII. Picnic tables had not yet arrived.

Production prototype Mark V, taken outside Wappenbury Hall, home of William Lyons.

shoes were 2¼in wide and operated in 12in cast drums. A normal cable-operated handbrake was employed. The 16in wheels wore the new Dunlop 6.70 Super Comfort tyres specially developed for the Jaguar to give silent running and longer life.

For the power unit Jaguar retained the Standard-designed 2½- and 3½- litre ohv straight-six engines, with milder camshafts for smoother running. The engines were mounted further forward in the chassis to provide extra passenger space. The smaller 1½-litre engine used in the previous models had now been dropped from the range, probably because the tooling was still owned by Standard-Triumph, who were using the engine in some of their own models.

Power was transmitted through the existing four-speed manual gearbox, which had single helical gears and needle roller bearings for intermediate gears and layshaft. The reverse shaft had been stiffened. A two-piece Hardy-Spicer propshaft passing through the chassis cross-bracing via a rubber mounting avoided transmission vibration and at the same time allowed a lower line to the transmission. This eliminated the usual transmission hump in the rear compartment. A Burgess exhaust system was fitted for both engine sizes.

Bodily the Mark V may have looked remarkably similar to the previous models but every single panel had been altered. Although retaining the same 10ft wheelbase as the previous models and being only 1in longer, the car looked much larger and was heavily adorned with chromework which may

The XK120 roadster was the first model fitted with XK engine, which, in various guises, was to power Jaguars up to the Series 3 XJ.

have suited the North American market but was not initially that well received in England. While traditional in style, the Mark V had a definite flow of line lacking in the older cars. Starting at the front, the most striking feature was double bumpers with heavy overriders, duplicated at the rear. Both bumper assemblies were made up of no less than thirteen separate components including mounting brackets. A substantial radiator grille was topped by a "real" plated filler cap that would also take the optional leaping Jaguar mascot if ordered. Faired-in headlights and twin fog/spot lights were standard equipment.

From the side there was a marked slope to the bonnet, and a more steeply raked windscreen than on previous models. The doors were wider, with plated window frames and bold plated waistline strips. Full spats covered the rear wheels and there were all-new push-button Wilmot-Breedon door locks (a feature really intended for, and then used on, the Mark VII and later models).

Rear end styling was similar to the older models, albeit with a larger rear window, heavy double bumpers, and the fuel filler cap hidden beneath a hinged panel. Inside the boot the same fitted tool kit was featured, continuing a theme from the pre-war cars.

A two-door drophead coupé version of the Mark V was also available and actually stayed in production slightly longer than the four-door saloon. Although basic construction was the same, extra bracing was provided by 9in wooden sections bolted to the chassis below the sills. Extra bracing was also used in the larger and heavier doors. Old style conventional door locks were used on the drophead coupés.

Inside, the Mark V was a traditional British luxury car, with liberal use of unpleated Vaumol leather upholstery on Dunlopillo cushioning, and walnut veneers on the dashboard and doors. The bucket seats had height adjust-

This early Press picture of the Mark V shows its sweeping lines.

In two-door drophead coupé form the Mark V was particularly appealing.

ment, a heater/demister unit was standard equipment, and there was a new means of dashboard illumination in the form of Lucas ultra-violet lights positioned under the dash top rail and reflecting off fluorescent instrument calibrations and pointers.

Mark Vs were available in up to 25 combinations of exterior and interior colour schemes, but despite popular belief the model was never offered in a two-tone finish as standard.

At launch prices of £1,246. 11s. 8d. for the 2½-litre and £1,263. 3s. 11d. for the 3½-litre versions the Mark Vs continued to enhance the Jaguar reputation for value.

As an interim model the Mark V had only a limited production run, with a total production of 14,500 of all versions. It gained a keen following, particularly for its advanced handling and utter reliability. It also proved an ideal test bed for Jaguar's new chassis and suspension, and a couple were fitted with the twin-camshaft XK engine for test purposes.

The next-generation Jaguar saloons in the forthcoming chapters owe a lot to it.

INTRODUCING THE MARK VII

Jaguar's launch brochure for the new Mark VII saloon read: "An entirely new car of unparalleled beauty".

While the Mark V should be considered a significant development for the Jaguar marque its replacement, the Mark VII ushered in an entirely new era for the firm. It was a car so fundamentally well designed and executed that it would influence saloon car design for many years to come. It subsequently proved to be one of the most important models ever to be produced by the Coventry company and clearly marked the transition from tradition to innovation.

It was always William Lyons' intention that Jaguar should make significant inroads into the lucrative luxury saloon car market: what he wanted to produce was a true 100mph saloon, with excellent performance both in acceleration and in top speed, of large proportions, not bulky but of imposing presence, capable of propelling five or six people in the utmost luxury and

The Mark VII frontal design taking shape with modified Mark V parts, this picture showing alternative lighting positions.

INTRODUCING THE MARK VII

This prototype, built around a Mark V, is very close to the final design for the rear, but with a lower front wing line.

Another front wing design for the Mark VII, reminiscent of the 1956 2.4 saloon.

comfort, and a car of true style to match the best available from Bentley, Lagonda or even Rolls-Royce. To coin a phrase thrust upon the public by Lyons on the introduction of the Mark V, a phrase that would be reiterated throughout production of the saloons covered in this book and epitomising the Lyons concept of a luxury saloon: "Grace... Space... and Pace".

The reader can be forgiven for thinking that we have perhaps missed a model out – whatever happened to the Mark VI? The story goes that after Jaguar's successful launch of the Mark V, Bentley brought out their new post-war saloon, designating it Mark VI. This meant that Jaguar literally had to "skip" a Mark to avoid confusing the public, and so the Mark VII was born.

Whilst the eager car-buying public had to wait until October 1950 for the launch of the Mark VII, the mechanics, including the engine, were virtually ready at the end of 1947. Due to the production methods required by the new bodyshell, Jaguar had to wait until tooling for the Mark VII had been completed, and production of the Mark V allowed them the breathing space they so badly needed. (The Mark V incidentally remained in production in 3½-litre form for a limited period after the launch of its successor.)

The Mark VII chassis complete with XK engine; note the upswept treatment at the rear.

We have largely covered the chassis layout and independent front suspension developed for the Mark VII but certain modifications were carried out for the new car. The front anti-roll bar had to be repositioned, with a revised steering column separated from a stronger steering box via rubber couplings to prevent vibration. These changes were necessary as the new engine was 3in longer than the old pushrod unit and was placed 5in further forward in the chassis to allow extra passenger space. This re-positioning of the engine also gave the new model better balance and thus improved its handling.

The same single-helical four-speed (three with synchromesh) Moss manual gearbox with short remote floor-mounted lever was used, with ratios of: 1st 3.375:1, 2nd 1.982:1, 3rd 1.367:1, 4th 1:1, reverse 3.375:1.

Power was transmitted via the divided propshaft to a now standardised Salisbury 2HA (later 4HA) rear axle with a revised ratio of 4.25:1. A single dry plate 10in Borg and Beck clutch was fitted.

For the first time on any SS or Jaguar the braking system was power assisted via a Clayton Dewandre servo situated under the driver's floor between the chassis crossmembers. The system was vacuum operated from the inlet manifold and worked only when the engine was running; immediately the engine was switched off the system became inoperative as there was no vacuum storage reservoir.

The Girling Autostatic hydraulic drum brakes at the front employed two trailing shoes instead of the more usual two leading shoes, and were self-adjusting. The rear drum brakes were substantially the same as on the Mark V, with one leading and one trailing shoe, manually adjusted. Both front and rear drums were the same size, 12in by 2¼in wide. The handbrake was repositioned between the front seats.

The XK power unit, developed for the Mark VII but initially released in the XK120 in 1948, deserves special attention as it was to become the mainstay of Jaguar production up to the introduction of the V12 engine in the E-Type of 1971, and still remained in mass production up to the demise of the Series III XJ6 in 1986. At the time of writing the engine is still in limited production for military use – an active life-span of forty-five years.

William Lyons had realised the limitations of the old pushrod 2½- and 3½-

litre units inherited from Standard in the 1930s, and although he bought the tooling for them in the immediate postwar period, he never intended to keep them in production for long. During World War Two he had discussed the development of an entirely new unit with his fellow colleagues and engineers William Heynes, Walter Hassan and Claude Baily. Through the long evenings of the war they had worked together on a design for a sophisticated new engine to power the next generation of Jaguar models through that magical 100mph barrier.

Because of the parameters laid down by Lyons the new engine had to be advanced in design yet easy to produce in large numbers. It had to supply sufficient torque throughout the whole rev range, be smooth and silent in operation and at the same time be tractable, reliable and easy to maintain. The engine also had to look pleasing and purposeful.

There were many attempts to arrive at the correct configuration, four, six, eight and even twelve cylinders being considered. Finally they produced a number of units (ironically under the "XJ" identification, the "X" standing for experimental) in the form of a four-cylinder twin-cam of 1,996cc (80.5mm × 98mm). This engine subsequently powered the Goldie Gardner record-breaking MG EX135 car, which achieved a maximum speed of 176.7mph in April 1948 with 146bhp from only two litres capacity!

While the four-cylinder engine was seriously considered for mass production, a larger six-cylinder version was needed for the smoothness and performance necessary to propel a luxury saloon to 100mph. So the "XJ" unit was reworked in six-cylinder form with bore and stroke of 83mm × 98mm, giving a capacity of 3.2 litres (3,182cc). This engine proved smoother but lacked the desired low-speed torque. A new configuration was arrived at by increasing the stroke to 106mm, giving a cubic capacity of 3,442cc, and with this the "XK" designation was born.

Twin overhead camshaft engines were rare at this time, usually reserved for expensive performance cars, whose owners put up with a lack of reliability, costly maintenance and other problems associated with complex high-performance units. So William Lyons' entry into the mass market with such an engine was well calculated to give Jaguar an edge over the competition.

The exquisitely designed twin-cam cylinder head for the XK engine, developed by Jaguar in collaboration with Harry Weslake, was made of high-tensile aluminium alloy for maximum lightness. Hemispherical combustion chambers were employed, with large-diameter valves set at 70 degrees in seatings of special high-expansion cast iron alloy shrunk into the head. Complete and efficient burning of the mixture was assured by the siting of the sparking plugs on the engine centre line.

The XK 3,442cc twin cam engine as originally fitted to the Mark VII saloon.

Weslake's participation encompassed the design of the specially contoured valve ports and induction system. The twin overhead camshafts (of the same lift as the XK120 sports but reprofiled to give better low speed power for the saloon) were driven by a two-stage timing chain and operated the valves directly through floating tappets, enabling very light valve springs to be used. Each camshaft ran in four shell bearings. The camshaft sprockets were carefully designed with fine adjustment in mind. Again through good design, the cylinder heads could easily be removed from the block without upsetting the valve timing. The camshafts and tappet faces were submerged in an oil bath formed in the cylinder head casting, allowing ample lubrication on all wearing surfaces.

The XK engine used a seven-bearing EN16 steel counterbalanced crankshaft, dynamically balanced, running in seven massive 2¾in Vandervell shell-type main bearings. Torsional vibration was limited by a Metalastic damper fitted on the front of the crankshaft. Steel connecting rods were employed, and the aluminium alloy pistons had chromium-plated top rings. Amazing rigidity was built into the XK crankshaft, ensuring ultra-smoothness of the engine throughout the rev range.

A gear-driven, large-capacity oil pump was driven off the front of the crankshaft, the oil passing through a full-flow filter into a ¾in gallery running the full length of the engine. The distributor was also driven off the forward end of the crankshaft. All this was within a cast iron block manufactured for Jaguar by Leyland. Twin 1¾in SU H6 carburettors were used and the new

The final design in the flesh.

engine developed no less than 160bhp at 5,200rpm on an 8:1 compression ratio (7:1 optional). Maximum torque of 195lb/ft was developed at 2,500rpm.

The whole engine was not only well designed internally but was also aesthetically appealing, with polished aluminium cam covers, chromium plated cam cover nuts, black vitreous enamelled exhaust manifolds and polished alloy inlet manifolds. The engine not only produced the required performance to reach 100mph, but also looked as if it did! The XK power unit was the heart of the Mark VII and must be considered one of the most significant engines in the motoring world. It is certainly the most significant engine ever built by Jaguar to this day.

Bodily the new Mark VII Jaguar was a masterpiece in styling, with superbly contoured curves. It was up-to-the-minute in its design and conception, yet still bore a traditional look of luxury and class, following a Jaguar style initiated with the XK120 sports.

The many large panels needed for the Mark VII body demanded gigantic presses and processes only available in a limited number of specialised outside bodybuilders. Lyons turned to the Pressed Steel Company in Oxford, who were to co-ordinate the Mark VII design, creating and assembling the

Jaguar press picture from 1950. Note the complex shaping of the bumper bar and the inset auxiliary lighting.

The rear-end styling minimised the bulk of the Mark VII.

body and delivering it to the Jaguar factory ready for paint preparation.

The Mark VII body design was a big step from the Mark V, but the Jaguar pedigree was evident in the long bonnet, the impressive frontal aspect, the plated window frames and the stylish sweep of the rear.

The front was imposing, with an abundance of chrome, Lucas 'tripod' headlights, flush-mounted fog lamps and sidelights in the tops of the front wings. The radiator grille, however, was far less dominant than on previous Jaguar saloons, losing the heavily plated surround and with a simple winged badge on the leading edge of the bonnet incorporating a simplified Jaguar head in place of the more imposing mascot. The bumpers, whilst less formidable than those of the Mark V, were still of complex ribbed design.

At 6ft 1in wide the Mark VII was 4in wider than the previous model although with its enveloping coachwork the extra size of the car was to some extent hidden. An unusual feature was the fitment of a split windscreen; unusual because Triplex had already developed the curved one-piece screen.

From the side the Mark VII was certainly a very impressive motor car, and though no less than 16ft 4½in in length it looked even longer, perhaps an illusion created by its relatively low overall height of 5ft 3in. Despite the stature of the new model there was a relative lack of adornment, with little brightwork except for the plated window frames and the bold waistline trim along the bonnet edges and doors, terminating in an "arrowhead" above the rear wings. The car had a rather slab-sided look but the sculpted wing line broke up the effect nicely, echoing the XK120. Full spats covered the rear wheels, held in place by twin Dzus fasteners.

Another first for Jaguar and many

other manufacturers of this period was the use of front-hinged doors for the driver and front seat passenger. The Wilmot-Breedon push-button door locks originally featured on the Mark V were used, and fresh air flaps in the sides of the front wings gave a cool air flow to the front footwells.

The same 16in Dunlop wheels and tyres were employed as on the Mark V, with the now familiar Jaguar-badged hubcaps, initially part-painted to match the body colour. Chromium plated Rimbellishers, held in place by screw and clip fittings, were also used as on the Mark V.

At the rear, the top-hinged boot lid had a chromium plated, square number plate housing, with attractive and practical twin chromium plated handles flanking the number plate. A scripted Jaguar badge adorned the centre of the boot lid.

Early cars had small circular rear lamps combining brake and tail lights, with separate round reflectors. A steel rear valance was hidden by the ribbed rear bumper, which featured prominent overriders. A single exhaust pipe exited below the valance on the nearside.

The Mark VII featured two independent fuel tanks mounted in the rear wings, their filler caps hidden behind flush fitting lockable lids in the tops of the wings. Each tank had its own fuel pump and fuel line. One tank held seven gallons and the other ten.

The boot itself was now 3ft 8in wide and no less than 4ft in length, accommodating an extraordinary 17 cubic feet of luggage. Jaguar claimed the boot could contain no less than four large suitcases, four sets of golf clubs, rugs, holdalls and other travelling sundries! With the fuel tanks in the wings and the tool kits in the front door panels, the only obstructions in the boot were the spare wheel, fitted upright in a well on the right, and the jack clipped to the floor. The boot interior, however, was scantily finished, with a Hardura covering to the floor and no spare wheel cover. The rather heavy boot lid had to be held in position by means of a hefty telescopic strut (sometimes prone to trip its stay catch).

Inside, the Mark VII was the most luxurious SS or Jaguar yet produced, providing armchair comfort to match any contemporary model, including the Best Car in the World. The re-positioning of the engine further forward in the chassis allowed an extra 2in of rear leg room. Along with this the rear seat was positioned further forward, providing extra boot space and giving the advantage that the rear wheel arches did not protrude into the rear seat.

The seats had Vaumol leather facings. The front bucket style seats were adjustable fore and aft and had height adjustment by means of chromium plated handles mounted on the front of the frame (like the Mark V). The 58in wide rear seat had a large fold-away centre armrest. To help access to the rear compartment the rear seat had rounded corners.

The door trims and seat side facings were upholstered in Rexine to match the grain of the leather. There were wide opening quarterlights front and rear, the latter with complex swivel hinges adapted from the Bentley Mark VI. The internal door handles were of an unusual trigger type (sometimes difficult to operate if the doors were incorrectly adjusted) and internal locking was by means of small plated circular catches on each door trim. The rear doors had childproof locks. All four doors had walnut veneered cappings. In each rear quarter were to be found matching courtesy lights for the rear seat passengers, and Mark VIIs featured no less than five ashtrays, amazing when one considers that William Lyons was an ardent non-smoker!

The tool kits in the front doors contained a comprehensive range of items including pliers, screwdriver, adjustable spanner, range of open-ended and box spanners, tyre pressure gauges, valve timing setting tool, bleeder tube in Lockheed tin, set of feeler gauges, grease gun, tyre valve extractor, spare bulbs, and spare spark plugs.

The flush fitting kits were released by pressing chromium plated button mechanisms which were, and still are, stiff to operate.

The carpets were of expensive Wilton with a leather gaiter around the floor-mounted gear lever. Wool headlining was used throughout and all Mark VIIs were equipped with a metal sliding sun roof as standard. The edge of the headlining was trimmed with veneer. The matching sun visors fitted flush into the headlining, the passenger one containing a vanity mirror.

The dashboard was traditional but stylish, with walnut veneer abounding. Instrumentation was extensive with a five-inch 120mph speedometer and matching rev counter, both with exquisitely figured digits and pointers. Between them were three smaller dials for fuel, amps and combined water temperature and oil pressure. A clock was set into the lower face of the rev counter. Auxiliary switches were mounted in vertical rows to the left and right hand of the instrument panel and included wiper, panel lights, map light, heater controls and cigar lighter. The heater unit itself, a very basic contemporary Smiths item although described as an "air conditioning unit", was merely a recirculating unit operated by a push/pull switch on the dashboard controlling the single-speed booster fan and a knurled knob varying the temperature from cold to warm.

Dashboard illumination was as before by low density ultra-violet bulbs mounted behind a veneered cant rail. On the passenger side of the cant rail would be found a substantial chromium plated grab handle, obviously to stabilize the passenger during the driver's spirited cornering. To either side of the dash centre panel were smaller veneered panels, the passenger side containing a lockable glove box (with inside illumination) and the driver's side a similar although smaller box with lid, both with baize interior coverings.

The Bluemels steering wheel was of a massive 18in diameter with adjustment for reach via a knurled ring. (A smaller 17in version was also available to special order). The massive centre boss contained a prominent horn push and the trafficator switch.

Underneath the dashboard would be found the bonnet catch, to the far right of the driver (right hand drive models), and fresh air ventilator levers in each footwell. A floor-mounted headlamp dipswitch was sited next to the clutch.

The Mark VII launch was a last-minute affair for the October 1950 Earls Court Motor Show in London, but what a launch it was. After the successful launch of the XK120 at the 1948

The facia of the Mark VII, with walnut veneered dashboard and giant Bluemels steering wheel. Note the prominent ashtray plinth on the top of the facia, later discarded in favour of twin ashtrays in the door panels. The pull-out veneered drawer was replaced by the radio, when fitted. Note also the footwell ventilators (later abandoned) and the crackle finish horn push surround (chromium plated on production cars).

The Mark VII prepares to wow the crowds on the eve of the 1950 Earls Court Motor Show.

A Jaguar training meeting to familiarise dealers with the Mark VII.

Motor Show the Mark VII (with the XK's engineering refinement) was the prima donna of the 1950 show, and to coin a phrase from the *Sunday Times*, "the Mark VII saloon undoubtedly stole the show". Perhaps as a thank you to his employees at Jaguar William Lyons arranged for them all to visit the Motor Show to admire the finished article and take in the public reaction.

With a launch price of only £988 (exactly the same as the 3½ litre Mark V which it replaced) and an all-in price inclusive of purchase tax of

£1,275.19s.6d. ($4,170 in America), Jaguar had done it again – in fact many said the car was too cheap, including some of Jaguar's own personnel, who strongly felt that even at a substantially higher price they could still have sold the car in great numbers. The *Daily Mail* said it all when it commented, "The car they all want. A world-beater if ever there was one..."

The UK release was followed by a similar launch for the American market at the Waldorf Astoria hotel in New York, where the car was displayed on a plinth set amongst satin and velvet. This was to be the first Jaguar specifically designed to capture the US market, by far the most important for a British car manufacturer then as now, and most of the initial production would be earmarked for North America. In fact orders worth over $3 million were taken in the first four months and a staggering 500 cars were sold in the first three days. Because of this success Lyons was able to negotiate with the Government to acquire the ex-Daimler wartime shadow factory in Browns Lane for increased production.

The Mark VII was available in a wide choice of colour schemes as standard although, as with any other luxury car of the period, special finishes could be ordered at extra cost. The standard range was based on the following:

Exterior Finish	Interior Trim
Suede Green	Suede green
Ivory	Red
	Pale Blue
Birch Grey	Red
	Grey
	Pale Blue
Battleship Grey	Red
	Grey
	Biscuit
Lavender Grey	Red
	Suede Green
	Pale Blue
Gunmetal	Red
	Grey
	Pale Blue
Black	Red
	Tan
	Grey
	Biscuit
Pastel Green	Suede Green
	Grey
Pastel Blue	Pale Blue
Dove Grey	Tan
	Biscuit
Twilight Blue	Blue

Some would say this was a rather sombre range of colour schemes, incorporating an astonishing assortment of greys, but there were also some interesting and classic colours like Twilight Blue, Pastel Blue and even Gunmetal, which was similar to a metallic charcoal. A few cars were supplied for export in a two-tone finish, usually a darker colour to the roof and pillars with a lighter colour for the rest of the body. These were never listed by Jaguar as standard options.

Contemporary competition

The Mark VII Jaguar was certainly launched at the right time. People were desperate to own new cars, but because of demand in the lucrative American market at which the car was specifically aimed, there was little hope of the poor old British getting their hands on one for some time!

This is not to say that competition wasn't strong in those days. The UK was still producing many fine luxury saloons from long established and reputable companies, all aiming at the same market. Leading the competition was Daimler, which was still an independent company. The 2½-litre Consort DB18 four-door saloon was similarly priced to the Jaguar but didn't have the performance or the panache, relying heavily on pre-war styling and engineering. By 1952 Daimler's new 3-litre Regency model offered good performance and modern styling but was by then much more expensive than the Jaguar.

Next in line stood Armstrong Siddeley with the Lancaster and Whitley saloons. More modern than the equivalent Daimler, these had an excellent reputation but were never very popular and certainly were not powerful cars even by the standards of the day. The company's new Sapphire 346 model of 1953, with its 3.4-litre 120bhp engine, was much more competitive, and was available with twin carburettors giving 100mph. Quite attractive and well appointed, the Sapphire could have given the Jaguar a run for its money if it had not been introduced so much later.

Another contender might have been the 2½-litre Lagonda, but it was very expensive, was produced in small numbers and never really had the flair of the Jaguar. It was later replaced by the more attractive 3-litre Tickford saloon, but this also suffered from being poor value for money.

Other upmarket competitors were Alvis and Bentley. The Alvis 3-litre TA21 was very much in the pre-war styling idiom, and though it handled well it was not as fast as the Jaguar and lacked interior space. The Bentley, in 4¼-litre Mark VI form, was a prestigious saloon and was an excellent driver's car, but at over three times the price of a Mark VII it was really in a different league. Even the later R-type version with 4½-litre engine and improved styling, while beating the Jaguar on build quality and presence, wasn't so much of a sporting saloon.

At the opposite end of the scale were the relatively mass-produced Humber and Rover models. The Humber Super Snipe had a very smooth 4-litre side valve six-cylinder power unit and was cheaper than the Jaguar, but again owed its styling to the pre-war period. The Mark IV which arrived in 1953 offered better performance, transatlantic looks and soon even a sporting reputation gained in the Alpine Rally, but by this time the Jaguar Mark VII was too well entrenched in the marketplace.

The Rover P4 offered six-cylinder twin-carburettor power with full-width styling, modern wrap-around windows and excellent quality at a cheaper price than Jaguar, but it was a relatively small and staid car, likely to appeal to a conservative taste.

The Jaguar, on the other hand, was squarely aimed at the man who was "making it in life". It was a fast car for a fast mover. It offered superb handling, impressive performance, and the appeal of a very stylish sports car, with boardroom comforts.

What the Press said

Although descriptions of the Mark VII appeared in magazines during 1950 it was to be 1952 before the Press got its hands on examples for test and long-term evaluation.

Gordon Wilkins wrote of his 6,000-mile experiences in *The Autocar* in February 1952. He felt the car's performance, road behaviour, appearance and finish made it one of the most impressive available in the world at the time – some accolade from a normally critical journalist. He did level some criticism at the servo braking system, which apparently allowed water vapour to condense in the servo cylinder and freeze in cold weather. With a rearrangement of the vents to the servo this problem went away.

Twelve months after the launch of the Mark VII, Jaguar's fortunes took another dramatic turn with the introduction of the XK120 "C" Type sports-racing car and its 1951 win in the Le

Mans 24-hour endurance race. This single race, more than any other sporting activity, put Jaguar on the map worldwide. From that moment Jaguar would be a household name in the automobile world.

Not only did this improve Jaguar's reputation but it confirmed the company's ability to produce superb engines and motor cars, and the Mark VII rode unashamedly on the back of this success. Jaguar started to take front cover colour advertising in *The Motor* and *The Autocar* depicting the Mark VII with the race winning "C" type.

After a 2,000 mile road assessment *The Motor*'s report gave the Mark VII the thumbs-up as "a high performance saloon car of exceptional all-round merit", and further, in the opinion of the test team, it "was one of the best cars submitted for road test in the post-war years". It is interesting to note that whilst they considered other cars of the period offered similar performance, luxury or space, only the Jaguar gave all these advantages in one package at a "quite moderate cost..." It is also interesting to recall that *The Motor* saw the Mark VII as a "European" car providing the handling and characteristics of a sound driver's car while at the same time being at home on the boulevards of America.

American success for the Mark VII was vital to Jaguar as an export earner, so *Road and Track* had the opportunity of testing the car thoroughly, involving several months with at least four examples of the model. This included over 1,000 miles from Los Angeles to Reno and back again. They initially complained of poor build quality, but later cars tested were apparently better in this respect. Minor irritations included the obtrusive steering wheel horn push, which got in the way of the driver when manoeuvering (later changed). They also felt the long reach of the gear lever from first to second was awkward, but got used to it. The first test car was draughty and subsequent cars also suffered from wind noise. However, *Road and Track*'s team were "sold" on the car and when they compared it dollar for dollar with the competition (American and otherwise) nothing could match it for value.

In 1953 *Auto Age* magazine completed an article "Jaguar Mark VII – Marvel or Myth", put together as a compilation of opinions from fifteen owners and mechanics. On ride quality one-fifth of owners surprisingly didn't like the Jaguar, finding it hard and too firm. Considering this was an American magazine interviewing American drivers, this judgement was not surprising as the home grown "barges" all had super-soft suspension. Most of the others, however, liked the stiffer ride and found it felt safer. On the handling side they couldn't fault the Jaguar, although some still preferred the wallowy handling they were used to on Chevrolets and Cadillacs; but most, after getting used to the Jaguar's fine, controllable steering, likened it to a luxurious sports car. It was therefore not surprising that some of the evaluators found the Jaguar a "driver's" car. On the whole, styling was found to be pleasing with the odd complaint about the bulbous looking boot. Performance also came out tops with the majority, who considered the Jaguar one of the fastest saloons around. Minor complaints concerned the horn push and wind noise (again) but all found the interior trim and finish exemplary. In conclusion all were happy with their "foreign" investment, which although not necessarily special value for money (costing the same as a Cadillac or Lincoln), offered a combination of fine quality, style, speed and handling, not all of which could be found together in one American product.

John Rundle with his immaculate original Mark VII, his first car and still owned by him to this day.

The Mark VII's American launch, on a bed of satin, at New York's Waldorf Astoria Hotel.

Hub caps on the Mark VII were partially painted to match the body colour of the car.

Clive Morris's lovely Mark VIIM with period white-wall tyres, seen regularly on the road throughout the year.

Both front doors of the Mark VII were fitted with tool kits. The opposite side to the one shown included spare bulbs and grease gun. Note the camshaft timing plate top left.

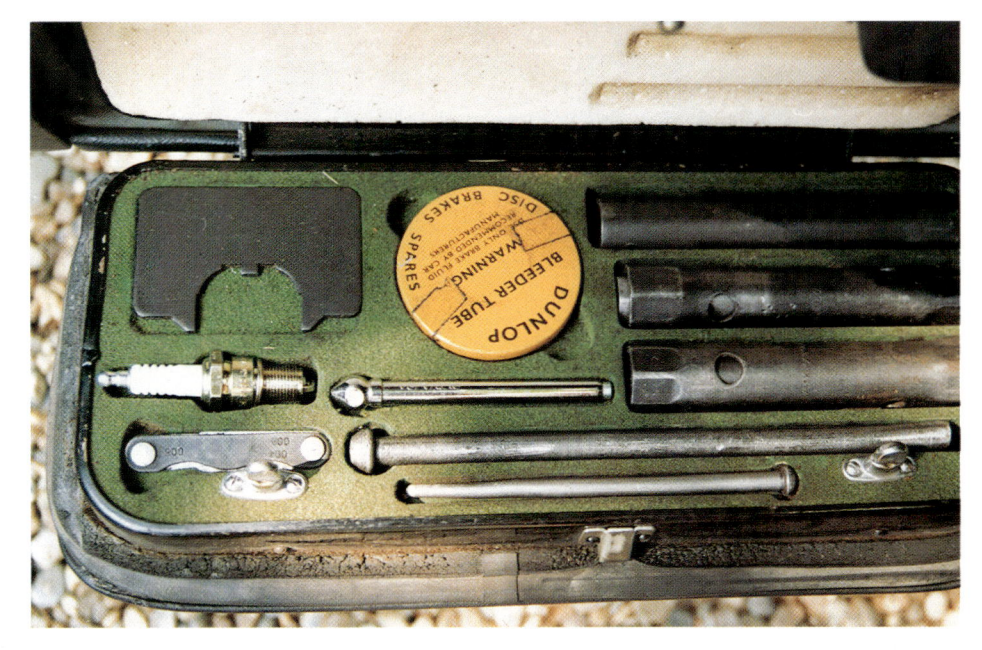

The rear bumper wraparound on the Mark VIIM is clearly visible. Note the gigantic overriders.

This superbly restored engine bay of a Mark VIIM shows many original features like the screenwash bottle and heater unit.

All in all, the Mark VII received a healthy press which would be drawn on by William Lyons many times in the early 1950s in advertising and brochures to enhance the prestige of the model. The original Mark VII brochure, a prestigious publication in its own right, featured no less than two pages of tributes quoted for the benefit of prospective owners, and it is worth repeating here just a few of those many Press accolades:

"The car they all want. A world-beater if ever there was one." *Daily Mail*.

"The great success of the Motor Show is, uncontestably, the new Mark VII Jaguar saloon. Its lines are remarkably modern and yet in impeccable good taste." *Le Monde* (Paris).

"Britain's No. 1 car... The Jaguar Mark VII is going to be a sensation at the Earls Court Show." *Daily Herald*.

"The hit of the Show, by all votes and by the sturdy number of orders taken, is the steel-blue Jaguar Mark VII sedan, an elegant job that can do a hundred plus miles an hour." *New Yorker*.

"The sensation of the British Automobile Show has been the Jaguar, a beautifully proportioned sedan of outstanding elegance." *Informaciones* (Madrid).

In the more recent past, Brian Palmer carried out a joint test in *Classic Cars* of March 1990 between a 1953 Mark VII and a 3.8-litre Mark 2 saloon. The Mark VII was an original, unrestored 1953 example having covered only 14,000 miles from new. Palmer found the Mark VII a memorable experience, emphasising that everything was on such a grand scale, reminiscent of the dignity of a dowager duchess! He particularly liked the car's excellent turn of speed and handling capabilities.

Italian motor racing ace Alberto Ascari bought a Mark VII in 1952.

Development of the Mark VII

As was and still is the case, constant development of models takes place during production to improve on the original concept, and the Mark VII Jaguar was certainly no exception to this rule.

1951

By the end of 1951 minor improvements were starting to filter through production. In November the thermostat assembly was changed, making parts interchangeable from engine no. A.2001. The cast aluminium radiator fan was replaced by a new steel type with redesigned pulley running at a higher speed to improve cooling. This also took effect from engine no. A.2001 but the modification was advised as an extra-cost option for previous engines using new part nos. C.5057 (fan) and C.5054 (pulley). Around the same time, from engine no. A.3303, mainly for cooler foreign climates, a boss was incorporated into the cylinder block at the rear left side of the engine above and slightly forward of the dipstick; the boss was fitted with an internal hexagon plug threaded for the fitment of an engine heater supplied by Electric Heating & Manufacturing Ltd.

At chassis nos. 710376 rhd and 730596 lhd a divided steering drop arm was fitted, on which the main drop arm was free to move on the steering rocker shaft, being operated from a secondary arm through a rubber coupling. Where this divided drop arm was fitted, the Ferobestos lower wishbone ball sockets were replaced by Morganite sockets, and the lower wishbone ball housings were fitted with grease nipples.

After many complaints (echoed in road test reports) of poor draught sealing and water leaks in the front doors, Jaguar advised the fitting of an extra rubber seal, part no. BD.2045/5, between the existing door seal rubber and the windscreen pillar and cant rail. It ran from immediately above the interior courtesy light switch, up the screen pillar and along the cant rail to the centre pillar. It was fitted with the lip of the new seal to the rear so that it bedded down against the existing sealing rubber.

1952

By January, at chassis nos. 711802 rhd and 732209 lhd, a two-speed wiper motor had been fitted to all models, with a new three-position dash-mounted switch, the switch identified with the markings P for park, N for normal speed and H for high speed. A Trico windscreen washer working by vacuum system had also been fitted, and was operated from a neat additional push button on the dashboard adjacent to the starter button.

In February, after complaints of excessive tappet noise, dealers were asked to amend the tappet clearances to .004in inlet and .006in exhaust.

At chassis nos. 712588 rhd and 732403 lhd, a new short-mainshaft gearbox without a rear extension was introduced, designated by the prefix letters SL or JL. With this modification a longer front propshaft was fitted, together with a revised speedometer cable of different length. It should be mentioned, however, that by April, due to supply problems, many Mark VIIs with later chassis numbers were fitted with the earlier long-mainshaft boxes.

At engine no. A.7027 additional fixing studs were fitted to the front of the cam covers to eliminate oil leaks from this area. Around this time, due to complaints of major oil leaks, Jaguar ascertained that a common fault was blockage of the engine breather pipe, which needed either cleaning out or re-positioning due to the pipe fouling the chassis; in the latter case Jaguar suggested cutting the pipe to a more suitable length.

In April uprated Girling shock absorbers were fitted as standard equipment on USA spec. cars.

In May, at chassis nos. 714860 rhd and 734987 lhd, wider and stronger 5½in Dunlop wheel rims were fitted. These wheels (ideal, incidentally, for radial tyres) were identified up to 1955 by two depressions in the rim adjacent to the valve hole, and thereafter by "5½ K" stamped in the well of the rim. The fitting of these wheels effectively increased the track to 4ft 8½in front and 4ft 10in rear. Tyres of up to 7.60 section could be accommodated on the new rims, and to quote a contemporary Jaguar Service Bulletin – "If oversize tyres are fitted, it is important to ensure that adequate clearance exists between the rear tyres and the bottom of the flange at the front of the wheel arches. The edge of the wheel arch can either be knocked forward or cut off to provide the necessary clearance". Just prior to these chassis numbers the Girling shock absorbers received stiffer settings for markets with rough road conditions.

In June Jaguar made available to special order a high-setting Smiths thermostat for use in countries where extreme cold conditions were experienced.

At engine no. A.6536 modified valve and tappet guides were fitted so that high lift camshafts (⅜in) could be specified if required without additional head modifications.

There had been a number of complaints about poor quality finish to interior wood veneer, leading to many dealers putting in claims for new parts to replace faulty items under warranty. Jaguar identified that most of the problems related to poor polishing, and suggested that local help was sought in repolishing as they could not guarantee to supply new parts to match the exact veneer or colouring of the woodwork in any particular car.

From October onwards Mark VIIs (from chassis nos. 733845 lhd and 714083 rhd) were fitted with new moulded rubber wiper blades (instead of laminated). These were part no. 741680, and were subsequently available as replacements for earlier cars.

Complaints were being made by owners about the auxiliary starting carburettor cutting out too soon (it should always cut out at 35 degrees C). Jaguar therefore advised all dealers to fit mod-

The Ace company produced wheeltrims for the Mark VII-IX models which are quite rare today.

Mark VIIs were adopted by some Police forces, particularly in the north of England and Scotland. This particular fleet was part of the Glasgow force.

ified thermostatic switches identified by a green spot on the front cover. This item became part no. C.2474.

Also in October, Jaguar became aware of supply problems with the original Newton and Girling front shock absorbers, and dealers were having to fit other proprietary makes. In some unfortunate cases these replacement dampers were longer, causing the suspension to become locked in the rebound position, with the resultant failure of the upper wishbone ball joints. All dealers were asked to check cars so fitted, ensuring the correct fully extended length measurement of 15 9/16in (39.5cms) between the centres of the mounting eye bushes.

From December synthetic enamel paint was adopted by Jaguar for exterior body finish, replacing the cellulose type previously used. This applied from body no. L.010744 (a few earlier bodies were also painted in synthetic enamel, numbers L.010567 to L.010614.) Synthetic paint was supplied by Pinchin Johnson and British Domolac. A one-pint tin was supplied free of charge by Jaguar with every new car built, situated inside the boot compartment behind the spare wheel.

With the change to synthetic paint came a revised range of exterior colours:

Suede Green
Birch Grey
Battleship Grey
Lavender Grey
Black
Pastel Blue (non-metallic)
Dove Grey
(all the above continued from the original range)

plus:
Cream (replacing Ivory)
Mediterranean Blue (replacing Twilight)
British Racing Green (replacing Pastel Green)
Pacific Blue (with blue or grey upholstery)

Also during this period, due to further complaints of poor draught sealing and scuttle vent water leaks, improved rubbers were fitted all round.

Up to this time Jaguar had supplied only one key, operating all locks on the car. From this period two were supplied, the traditional round-headed key for doors, ignition and petrol fillers, and a rectangular-headed key which operated the glove box and boot.

At the very end of 1952, Jaguar expressed concern at comments by dealers to owners about the need to decoke engines every 4000-6000 miles, which was obviously incorrect. Jaguar believed the XK engine was capable of between 30,000 and 40,000 miles before decoking.

1953

After two years in production it was inevitable that some Mark VIIs would be involved in front end collisions, and with this in mind Jaguar started to issue separate part numbers for the components of the radiator grille, eliminating the need to replace the whole unit; subsequently every single slat became available under an individual part number.

In February a modification was introduced from engine no. B.2917 incorporating an eight-bladed cooling fan, a narrower fanbelt, and a water pump with Hoffman bearing, carbon seal and enlarged water bypass. A modified induction manifold was included. The grease nipples originally fitted for lubrication of the fan and water pump bearings were no longer required, the pump spindle bearing being prepacked with lubricant. Prior to this modification, at chassis nos. 715261 rhd and 735341 lhd, a modified type of radiator (part no. C.7613) was fitted, incorporating expansion chambers at the top of the header tank, one on each side of the filler cap. The position of the drain tap remote control rod was revised.

In March changes were made to the front brake wheel cylinders with the adoption of aluminium instead of bakelite air excluders (interchangeable). In the same month all 7:1 and 8:1 compression engines received new distributors incorporating high-lift contact breakers. The 8:1 compression engines also changed to SR carburettor needles.

In April, due to the odd complaint of brake servos grounding on rough terrain, Jaguar started to raise the servo on models from chassis no. 715582 rhd and 735920 lhd, and fitted it with a protective undershield on the bottom face of the chassis.

Also in April Jaguar started to make up Continental Touring Kits for the Mark VII containing all spares necessary in the case of minor breakdown. The kit was very well presented in a wooden box with Jaguar logo and full instructions, and was available on a sale or return basis, the hirer being charged upon return for those items used, if any. The price was a princely £24.0s.3d. plus an extra 6s.3d. for the tax on the bulbs!

In May, from engine no. B.4384, a malleable iron crankshaft vibration damper was fitted instead of the cast iron one. During the same period Jaguar reversed the fitting of rubber rebound stops attached to the bracket on the outside of the chassis. The thicker end of the stop then fitted towards the inside of the chassis side member. Jaguar recommended the same reversal for existing cars when service work was due on steering or suspension.

From chassis nos. 715749 rhd and 736143 lhd, all Mark VIIs were fitted with a 4HA Salisbury rear axle (part no. C.7385) in place of the previous 2HA version. The later axles are identifiable by the casting number embossed on the left hand lower web. The axle ratio of 4.27:1 remained unchanged.

The big change to the Mark VII specification list came in March 1953 when automatic transmission became an option on export models. Automatic transmission was the norm in the valued North American market, with over 90% of new cars supplied so fitted. At the time only one other English manufacturer, Rolls-Royce, offered this type of transmission option on export models, and Lyons saw the opportunity to enhance the Mark VII specification and therefore sales in this much valued market place.

Several types of automatic transmission were tested before the final choice of Borg Warner was made, this particular model being ideally suited to the XK engine. Imported initially from America, the unit was later to be made in England and became the preferred choice of two-pedal transmission for many other British models.

Based on a design used by Studebaker, the transmission used a single-stage torque converter with a ratio of 2.15:1 and a double-train epicyclic gearbox giving two forward speeds plus reverse. The low gear ratio was 2.308:1 and the intermediate gear 1.345:1. With a rear axle ratio of 4.27:1 the lowest overall ratio was 21.2:1 in low gear range, rising to 9.86:1, and in intermediate range from 13.2:1 to 6.14:1. At that point the gear train was disconnected and the torque converter locked, giving a direct drive of 4.27:1.

The ratios were selected by brake bands and a multi-disc clutch. A plate clutch, mounted just behind the engine, locked up the torque converter for top gear. These were all operated hydraulically by oil pressure generated by two pumps, one driven from the input shaft of the transmission, the other from the output side.

As well as automatic selection of the gears via pressure and acceleration of the car, a manual selector was fitted, operated from a dash-mounted control linkage, rather like a conventional column gear change. The selector lever and its illuminated Bakelite quadrant were sited above the steering column rather than integral with it.

All automatic transmission models featured a bench front seat, which meant moving the handbrake, an umbrella type being fitted to the underside of the dashboard nearest the driver's door. This feature was later standardised for all models.

The automatic transmission selector lever featured positions P-N-D-L-R. The system was planned so that the car could only be started with the lever set to P or N.

With the lever set to D, upon acceleration the torque converter ratio would automatically be reduced until it was operating as a coupling. The next stage was for the direct-drive clutch to lock up the converter. This would take place at anything between 18 and 55mph depending on the throttle opening.

At speeds of below 60mph it was possible to re-engage the intermediate gear by means of a kickdown facility. Direct drive would be re-engaged upon releasing the accelerator or when a maximum speed of 68mph had been reached.

Similarly L could not be manually selected at excessive speed, nor could R or P be engaged when travelling forwards at more than 10mph.

The system, although archaic by today's standards, was very acceptable at the time and certainly on a par with other automatic transmissions on offer. It enabled reasonably quick standing starts, virtually unaltered top speeds and even engine braking on downhill stretches by appropriate use of the low gear ratio. The new model also incorporated an ingenious anti-creep device via brake line pressure being applied to the rear brakes when Drive was selected, preventing the car from moving forward until the accelerator was pressed, at which time the brake pressure was released. All automatic transmission models of the Mark VII received the suffix BW after the chassis number.

In April Jaguar released information

on an engine modification to increase performance of the XK power unit by the use of an improved C-type cylinder head with revised porting, 1⅝in diameter exhaust valves and enlarged valve throats. This cylinder head came complete with valves, springs, high lift camshafts and new rocker covers on an outright sale basis (no exchange) for £150. 2in SU carburettors were also available at £18 each, requiring the fitment of a revised inlet manifold at £6.13s.

In June, from engine number B.5305, a pressed steel sump unit replaced the previous cast alloy type. The new sump meant a change in oil capacity to 19 pints (22 for the engine in total) and a new dipstick with the high level mark 9in from the combined stop and felt oil retainer (8¾₆in on the aluminium-sump dipstick). The oil pump incorporated a non-adjustable oil pressure relief valve and new oil filter. Coinciding with the above changes new Ferodo DM.8 bonded brake linings were fitted. Also, from chassis nos. 716412 rhd and 736656 lhd, twin throttle return springs were fitted, allowing more accurate slow-running settings. These were fitted to levers on the carburettor spindles and to brackets attached to the lower carburettor flange studs.

By August Jaguar had adopted the use of telescopic rear shock absorbers, from chassis nos. 717190 rhd and 736872 lhd. To accommodate these Jaguar had to modify the chassis with an extra cross member for the top mountings and a longer flexible brake hose bracket. The bodyshell had two holes drilled, with removable covers, at the front end of the luggage boot, to provide access to the top mountings of the dampers. Interchangeability of the earlier rear axles was no longer possible as the lugs fitted to accommodate the old style piston shock absorbers would be fouled by the telescopic dampers and possibly the flexible brake hose.

Due to some complaints about heavy clutches Jaguar fitted an overcentre spring and link between the clutch pedal and master cylinder bracket, thus reducing the effort required for operation.

At this time the company decided to fit new horns to the Mark VII with a more strident and penetrating tone. They were positioned in front of the radiator instead of the previous position on the wing valances. The new horns did not need the relay previously required, and were fitted from chassis nos. 716017 rhd and 736502 lhd.

Another change in August, to automatic transmission models from chassis no. 736871 BW, involved fitting a direct drive pawl to the extension case assembly, allowing a downshift from direct drive to intermediate gear to be made at speeds below approximately 45mph under conditions of heavy acceleration short of kickdown.

By September Jaguar had changed the specification of the rear springs on all Mark VIIs, from chassis nos. 716941 rhd and 736805 lhd, with an eight-leaf version (part no. c.7914) as opposed to the older nine-leaf type.

Also at this time torsion bar settings were amended to obtain a height of 8in between the point where the front end of the cruciform member meets the chassis side member, and the ground.

In the same month a new, hotter cigarette lighter was fitted, identifiable by 4 indents on the chromium-plated shoulder and a copper-plated ejection spring to the element. These later types of lighter burn their elements out immediately if used on earlier cars.

In November Jaguar issued a directive concerning complaints of brake judder. In such cases dealers were to reduce the length of front brake linings by cutting off around 2½in from the trailing end of each front brake lining.

By this time the basic cost of the Mark VII manual saloon had risen to £1,088 plus purchase tax of £605. 18s 11d. making a total of £1,693. 18s. 11d.

At the end of 1953 the colour range available on Mark VIIs had again been revised and rationalised:

Paint	Interior
Black	Red, Tan or Grey
Old English White	Red or Pale Blue
Lavender Grey	Red, Suede Green or Pale Blue
Battleship Grey	Red, Grey or Biscuit
Suede Green	Suede Green or Tan
Dove Grey	Tan or Biscuit
British Racing Green	Tan or Suede Green
Pacific Blue	Blue or Grey
Birch Grey	Red, Grey or Pale Blue
Pastel Blue	Blue
Indigo Blue	Blue or Grey
Claret	Red

1954

Changes to the Mark VII started with an amended chassis, fitted from nos. 718919 rhd and 737192 lhd onwards. The new chassis had front suspension posts of increased height. All such chassis frames were stamped with the letters SC. By May a uniform type of chassis frame was adopted as a "Service Condition" type adaptable for all modifications.

In January Jaguar fitted twin drain tubes to the carburettor air intake pipe to avoid spillage of excess fuel in the engine bay that could result in petrol fumes entering the car. This took effect from chassis nos. 719155 rhd and 737256 lhd, but many earlier cars were subsequently modified.

At chassis nos. 719362 and 737389 a new heater system was installed, with wider diameter pipes to improve flow. At the same time Jaguar made available a remotely controlled radio aerial. This involved a handle fitted to the scuttle side casing on the driver's side, immediately behind the front wing ventilation aperture. The handle, when turned, operated a pulley and cable system which raised or lowered the aerial.

Also in January a further transmission option arrived in the form of a Laycock de Normanville overdrive system operated by means of a hydraulically controlled epicyclic gear on the output shaft of the gearbox. Overdrive operation was semi-automatic. It was controlled by means of a fascia-mounted perspex switch situated on the top cant rail of the dash to the right of the steering wheel. When the switch was pushed to the right the overdrive was cut out. When pushed to the left it electrically triggered the overdrive to the "in" position, whereupon it only engaged above 36mph in top gear. If the speed dropped below this figure the overdrive would automatically trip "out" upon touching the accelerator. When in operation the overdrive switch was illuminated.

To accommodate the overdrive a Salisbury 2HA rear axle with 4.55:1 ratio was fitted, giving an overall ratio in overdrive of 3.538:1. Incidentally, from chassis nos. 723447 rhd and 738259 lhd the 4HA type Salisbury axle was fitted in place of the 2HA although retaining the 4.55:1 ratio. A revised speedometer was also fitted to suit the change. The overdrive produced lower engine revs, quieter running and much more economical high speed cruising. In fact, a 17% saving in engine revs could be achieved in top gear. A speed of 100mph in overdrive represented 4,292rpm as against 5,520rpm on the non-overdrive model. The overdrive, operating on top gear only, cost a very modest £45.

In April modified inlet and exhaust valves were fitted from engine nos. D.1079 (inlet) and D.1167 (exhaust) but including number D.1132. The modified inlet valves, part no. C.8248, were identified by a depression in the valve head, and the exhaust valves, part no. C.8313 (with accompanying guides), by a reduced diameter stem under the valve head to form a scraping edge inside the guide.

In May, from chassis nos. 720783 rhd and 737728 lhd, a front bump stop bracket assembly was fitted to the lower wishbone to provide a larger area of contact for the bump rubber. Because of complaints of fumes inside the car the petrol tanks received attention from chassis nos. 72107 rhd and 737770 lhd. The petrol filler boxes were moved rearwards and the lids lengthened and hinged at the rear instead of the front as previously. All this necessitated the modification of the petrol tanks, in future to be part numbers BD.9196 left hand and BD.9197 right hand.

At engine number D.1575 onwards a modified chain tensioner blade, hard chrome plated and polished, was fitted. For 7:1 compression engines SR carburettor needles were now fitted, in line with 8:1 compression engines.

By the end of July the Mark VII had increased in price slightly again to £1,140 (£50 dearer before tax) without overdrive.

A major development came in September 1954, with the introduction of a new version of the Mark VII to be known as the Mark VIIM. The changes were relatively significant and included for the first time substantial alterations to the external styling, although the

The dash-mounted overdrive switch. The same type of switch would also be used for the Intermediate Speed Hold on automatic models.

A late Mark VIIM with the ventilator flaps in the front wings deleted.

more practical improvements were in performance and handling.

From the side little looked changed except that both front and rear bumpers had wraparound sections to protect the corners of the body. The bumpers themselves were entirely new, with less pronounced shaping (echoed in subsequent XK sports cars and the Mark I saloons), accompanied by much deeper overriders for better overall protection. The rear bumper was mated up to the bodywork with a rubber strip in between. Wheel Rimbellishers were now a standard fitment on all Mark VIIs, previously an extra cost option.

At the rear, apart from the new bumper arrangement, the rear lights were replaced by the slightly larger Lucas L.549 type incorporating reflectors and prominent chromium plated mounts. (Some earlier Mark VIIs had been fitted with this new type lighting before the introduction of the Mark VIIM). Badging remained the same on the boot lid and did not incorporate the "M" identification at all.

At the front other changes were being made. Whilst the grille, badging and shaping of the panels remained, the lighting was substantially altered. Round flashing direction indicators were fitted, replacing the previous semaphore type. They were mounted about 6in up from the bumper level in each front wing. Side lights remained on top of the wings. The headlamps were changed for the more up to date Lucas J700 units with heavier chromium plated surrounds and Le Mans-type diffuser lenses. The matching fog lamps were also changed and moved from the usual flush fitting position to pod mounts on the front valance, allowing extra adjustment. These new Lucas SFT 576 units were common to many other luxury cars of the time, including Rolls-Royces. New chromium plated horn grilles were fitted where the fog lamps had previously been. Along with the new bumper treatment the frontal aspect was much more impressive and pleasing to the eye, and would remain virtually unchanged throughout the lifespan of the body style to 1961.

Inside, new and much deeper Dunlopillo seating was fitted, still with separate front seats on manual transmission cars and a bench seat for automatic models. Bench seats did not have picnic tables fitted in the back as is often thought (this only arriving on Mark VIII and IX models). All Mark VIIM models were now fitted with a revised steering wheel with flat horn push instead of the cone type used before. The front door trim panels now had long map pockets which in practice, although being deep, would not accommodate very thick maps.

The "new" Mark VII took its "M" designation from the "M" engine derived from the XK120, having high lift ($\frac{3}{8}$in) camshafts as standard compared to the normal $\frac{5}{16}$in type. These camshafts had previously been an extra cost option. They allowed the engine to run more efficiently at higher revs, and along with a new Lucas ignition coil increased the maximum power from 160bhp to 190bhp at 5,500rpm.

A close-ratio gearbox was also fitted which, with the standard 4.27:1 rear axle ratio, gave gear ratios of 12.73:1,

Stylish Jaguar bootlid script on the Mark VII.

7.47:1, 5.16:1 and 4.27:1. With the 4.55:1 rear axle of the overdrive model, ratios were accordingly 13.56:1, 7.96:1, 5.5:1 and 4.545:1, going down to 3.538:1 in overdrive top.

To accompany the performance modifications the torsion bars were increased in size to 1in diameter from $^{15}/_{16}$in, giving a higher spring rate.

Export Mark VIIM models featured Dunlop tubeless 6.70 × 16 tyres, whitewall versions being optional. Wing mirrors were also standard equipment on export models.

Prices for the upgraded Mark VIIM model in overdrive form were £1,185 plus £494. 17s 6d. in purchase tax making a total of £1,679. 17s. 6d, unchanged from the previous model.

A wide range of optional extras was available for the Mark VII cars, including:

Oil bath air cleaner system
7:1 or 9:1 pistons
High-speed tested crankshaft damper
"C" Type performance cylinder head with two 2in SU carburettors
XK120-style straight through exhaust system
Lead bronze big end bearings
High-ratio steering box
Various tyre options including whitewalls
Borg Warner automatic transmission
Laycock de Normanville overdrive (manual transmission cars only)
7.750 × 16 oversize Dunlop tyres (necessitating rear wheelarch modifications)
Wing mirrors
Seat covers
Witter towbar equipment
Three options of radio set including a 5-pushbutton HMV
Non-standard exterior paint options to suit individual taste

With the Mark VIIM came the availability of specially made and fitted luggage for the boot, made by Auto-Luggage Limited in Essex and marketed through the Jaguar dealership Henlys of London. Two fitted suitcases neatly fitted the irregular shape of the Mark VII boot and were sold for £45. 12s. 9d.

The Mark VIIM had a revised frontal treatment, with separate Lucas fog/spot lights, dummy horn grilles, and separate indicator lights eliminating the trafficators.

They could be had in a variety of colour schemes to match the car interior or exterior.

Interestingly, modified rear spats became available, with a cutaway section at the bottom, similar to those that would later appear on the Mark VIII saloons.

There were some other changes in 1954. Jaguar amended the specification of rev counters used on Mark VIIs by the adoption of part no. C.9183, from chassis nos. 724688 rhd and 758528 lhd. The red danger banding now spread from 5,500 to 6,000rpm instead of 5,200 to 5,500rpm previously. Another gauge change affected the combined oil pressure/water temperature unit, which was superseded by part no. C.9542 reading up to 100lb sq/in instead of 60lb. Finally, diecast metal replaced brass in the production of the radiator grille.

1955

Even after the introduction of the Mark VIIM, modifications continued in abundance, and in February 1955 the front wing ventilator opening flaps were eliminated, rendering parts for the older wings obsolete. This meant that any accident damage to older cars' wings would necessitate the fitting of new style wings without the ventilators. At the same time the body parts for the centre B/C post on earlier cars with trafficators became obsolete.

During the same period, probably due to shortage in supply, 2HA Salisbury rear axles were again substituted on right hand drive overdrive cars from chassis nos. 724365 to 724415.

Also in February, only very shortly after the start of production of the Mark VIIM, the clear indicator lights in the front wings were changed for the later Lucas L.563 type, which was to be used on many other Jaguar models including the contemporary XK140. It had an amber lens, although a clear lens was available for certain export markets.

Around this time Jaguar started to phase out the painted hub caps in favour of plain chromium plated versions of exactly the same design.

All export models had drilled front wings (with rubber grommets fitted) to accommodate the wing mirrors supplied as standard, although these were not fitted until P.D.I. checks were carried out by the dealers.

From engine no. D.6723 onwards a rotor type oil pump was fitted. In such cases the oil pressure relief valve was fitted to the filter head. For 7:1 compression engines high lift (⅜in) camshafts were fitted.

By March Jaguar had adopted Dunlop Tubeless tyres for all Mark VIIs. Also that month Jaguar fitted a modified type of 4HA rear axle which had the road wheels positioned further inboard of the rear hub bearings. This was effected by repositioning the brake drum flanges on the rear hubs and dishing the backplates to suit. To maintain the track of the rear wheels the overall width of the carrier and tube assembly was increased and the half shafts were lengthened. This modification took place from chassis nos. 725634 rhd and 738836 lhd.

For production of all Jaguar engines from September 1955, and in the case of Mark VIIs from engine no. D.9869, a spring-loaded Renold tensioner was fitted in place of the spring blade tensioner on the lower timing chain. A nylon damper assembly was also fitted in the position originally occupied by the guide bracket for the spring blade tensioner. The cylinder block had to be modified with an oil feed to the timing chain via the tensioner.

In December Jaguar also modified the cylinder heads for all models by reducing the depth of the tapped holes to accommodate the studs on the inlet face of the head. This affected Mark VII engines from no. N.2497. Slightly later, from engine no. N.2942, a modified type of oil filter was fitted, identified by the head of the centre bolt being at the bottom of the filter canister and not at the top of the filter head as previously.

From engine no. N.3482 the mounting of the torque converter was altered. The six mounting studs were replaced by tapped holes and the converter was secured to the engine drive plate by setscrews and tab-washers. The drive plate was increased in size to 11 3/16 in and the starter pinion housing cutaway was increased to avoid fouling the engine drive plate.

1956

In January the first automatic transmission version of the Mark VIIM became available on the home market.

In April Jaguar added Imperial Maroon to the range of exterior paint finishes on Mark VIIs.

The Mark VIIM model remained in production up to July 1957, by which time a total of 7245 right hand drive and 1688 left hand drive examples had been built on top of the 20,939 Mark VIIs, making a grand total of 29,872 over a period of six and a half years. By this time the top of the range automatic transmission Mark VII had risen in price to £1,268 plus £635. 7s purchase tax, a total of £1,903. 7s. The Mark VIIM actually continued in production for about nine months after the introduction of its successor the Mark VIII.

Later Mark VIIs: what the Press said

As we have already heard in the last chapter the Mark VII was very well received by the world's Press, and the introduction of automatic transmission, followed by an overdrive version and then the Mark VIIM, gave the motoring media yet more to enthuse over.

On September 17th 1954 *The Autocar* was the first to test the Mark VII with overdrive. The test was to some extent more critical than in previous reviews of the model earlier in its life. Nevertheless the magazine found the car superb overall, with a three-figure top speed and a cruising speed of around 80mph in overdrive, this speed governed only by the road conditions in England during the test.

Performance was measured as:
0-50mph 9.9 secs
0-60mph 13.6secs
0-90mph 35.8secs
Standing quarter mile 19.3 secs
Top speed 102.1mph, with a best recorded of 104mph

True top speed and economy were little different from the non-overdrive model but the car cruised with greater ease at higher speeds. *The Autocar* did, however, complain of the length of clutch travel and the long lever movement from first to second gears, and felt that although the ratios were well chosen the synchromesh could easily be beaten. Here again such criticisms could easily have been levelled at the earlier model as well.

The overdrive unit was judged excellent, with particularly smooth changes up and down. The tester also found that with overdrive engaged the car travelled so smoothly that both driver and passengers had the feeling of travelling much more slowly than they actually were.

In January 1955 *Wheels* magazine carried out a full road test analysis of the Mark VII with automatic transmission and headed the article "A definite advance in motor car design". On a 35-mile exploratory trip to test the car they found no need to move the transmission selector lever from "D" as the car coped happily with normal driving conditions and a 1 in 10 gradient. They also commented that on test the intermediate gear could easily cope with a 1 in 5 gradient, but felt that the intermediate ratio came in too early when accelerating hard. All in all it was an excellent report, though there was adverse comment on the car's braking ability at high speed.

Interestingly, (particularly in view of the fact that this was an Australian car magazine report), *Wheels* enlisted comments from women drivers on the Mark VII, who found the Jaguar large and powerful but very responsive and simple to control. They all liked the automatic transmission, which allowed them to concentrate on driving without the need to worry about gear changes. An interesting comment from one lady, indicative of this period when cars were quite basic in their equipment specification, was, "It is certainly luxurious. It has a heater and a sunshine roof...". How things have changed.

September 1955 saw the first road test of the Mark VIIM model, carried out by *The Motor*, which described it as "An exceptionally successful car in improved form". Introducing their report, they commented, "A small number of motorcars have so powerful a personality that they are able to impress the road tester with a key word to represent their particular quality. Such a one is the latest M type Mark VII Jaguar; the word which it conjures up is 'exceptional' ". *The Motor* seems to have had nothing but praise for the Mark VIIM in overdrive form, commenting on the exceptional quietness and tractability of the vehicle, the engine pulling smoothly from as little as 500rpm. The new versions of the Mark VII were found to be slightly more economical and still outstanding value for money.

Performance figures had hardly changed from the earlier Mark VII tested, although the 0-90mph time had been brought down from 35.8 secs to 33.4 (but this could have been due to the individual cars tested and to road conditions). Fuel consumption with the benefit of overdrive was 25.5mpg at a steady 50mph, with an overall average of 18.8mpg.

The Autocar followed up with its road test of the automatic Mark VII in 1956 and the accolades to Jaguar flowed freely. The tester found the automatic car down on performance compared with the earlier road test of a manual version, losing 0.3 secs on the standing quarter mile and 1.1 sec from rest to 80mph. The anti-creep device was found a little annoying in busy London traffic when manouvering.

Also in 1956 John Bolster, for *Autosport*, had the opportunity to road test the Monte Carlo-winning Mark VII, registered PWK700, which had been on display in Henlys showrooms in London. Driven by Ronnie Adams, the car had been campaigned in two Monte Carlo rallies as well as a production car race at Silverstone, but apart from some necessary trim and accessory modifications was generally to standard specification. A specially raised bucket seat was fitted for the driver and a headrest for the front seat passenger/navigator; extra padding was also provided along with the usual rally instrumentation, fire extinguisher and the like. The weather conditions for the road test were typically Monte, with heavy snow showers and ice.

Initially, driving through the heavy London traffic revealed a car of massive proportions which was heavy to manoeuvre, but when on the open road the car appeared to diminish in size. It behaved impeccably. Even when badly snowdrifted side roads were encountered the Mark VII stormed though in second gear, its sheer weight carrying it through where other cars couldn't go.

Moving on to a snowy Brands Hatch circuit John Bolster took the Mark VII onto the track, obviously enjoying himself. "...Wheelspin all the way, old man...", he said. He found the car could be handled quite safely on the limit, and even found the brakes excellent, without fade, although the steering was particularly heavy due to the fitment of special Dunlop extra-grip winter tyres all round. All in all Bolster voted the Mark VII an excellent winter rally car as well as superb everyday transport for VIPs.

Even in more modern times the Mark VII has held its own on road test. In July 1981 Paul Skilleter of P*ractical Classics* carried out a comparison test between a Mark VIIM and a 1950 Cadillac Series 62 with a 5.4-litre V8 engine. The choice of the Cadillac was deliberate, as it was typical of the US luxury cars at the time and came from one of the most reputable marques in America; the car was also new in concept and offered similar capabilities to the Jaguar. Although $647 cheaper than the Jaguar, the Cadillac was found by Skilleter to have better build quality and a superior heater, and to be easier to drive. The Jaguar, however, handled better and was quieter and faster.

The Mark VII today

Since the golden era of the Mark VII in the 1950s the car has taken very much of a back seat, particularly as it did not have the aura of the Jaguar sports cars and was very much overshadowed by the success of the 1960s Jaguar saloons.

With the growth of the classic car movement fortunes changed for the Mark VII, although prices remained stable, encouraging a small band of enthusiasts to run and care for the cars. Many classic car magazines in the late 1970s and 1980s featured the model, and Richard Sutton of *Classic and Sportscar* actually ran a Mark VIIM as everyday transport, finding the car a joy to drive.

Whilst working in London Clive Morris noticed Richard Sutton's Mark VIIM for sale. Although he had no experience of Jaguars at all and was really looking for a 2.5-litre V8 Daimler at the time, Clive knew of the Mark VIIM via the articles in *Classic and Sportscar*. After three hours deliberation and without so much as a test drive Clive took the plunge and purchased the car.

The car is an M model, without overdrive, registered in 1954. Although it has had a total of nine known owners Clive considers the indicated 54,000 miles could possibly be genuine. Finished in Lavender Grey with red interior, the car is still virtually original in specification except for the fitment of a 1960s Rover P5 brake servo, which seems to improve the performance of the drum brakes no end.

On a Saturday morning drive through Clive's home town of Leicester the car proved particularly nimble and practical. With an excellent turn of speed, superb brakes and the slightly throaty rasp of the exhaust, this Mark VIIM really did explain the success of the model.

When compared with other cars of the early 1950s there is no doubt that the Jaguar was the best available, and it is still a very usable car today. Clive regularly uses the car to drive to and from work including a section of M1 motorway where the car sits happily at between 70 and 100mph without trouble, although Clive does admit that the fitment of overdrive would make it even more enjoyable. Even so he still maintains an average 15mpg under all conditions of driving.

Clive has fitted Goodrich American whitewall tyres, which look particularly attractive set against the Lavender Grey paintwork and don't look out of place (particularly as most cars destined for the North American market were so fitted from new), but whilst the car handles well in the dry it slides badly in the wet particularly across road markings.

Clive's Mark VIIM is not a fair-weather car as he enjoys driving it so much. The sunroof unfortunately leaks a little so Clive tapes it up during the winter to avoid problems inside the roof. The car has had extensive work carried out to remedy rust and rot in the usual places, with some rechroming. The engine was rebuilt due to high oil consumption and numerous leaks; the opportunity was taken for a complete engine bay tidy-up and a suspension rebuild including fitment of Koni adjustable shock absorbers.

As an aircraft engineer Clive appreciates how well the Mark VIIM is built and finds it easy to maintain, without the complexities of later classics. He is Registrar for the model in the Jaguar Drivers Club and an active member of the Jaguar Enthusiasts' Club.

Next was a trip to Scunthorpe to meet John Rundle, another keen owner, this time of the earlier pre-M model. A late manual car with overdrive, it is resplendent in black with red interior, and dates from February 1954.

John's situation is perhaps unique in that the Mark VII was his first car, bought for him by his father in 1967. The history of 280BMT is unknown but John thinks that it has covered 75,000 miles although he has changed the speedometer twice in the last 25 years through necessity.

Although his Mark VII is now relegated to summer use amounting to about 1,000 miles a year, John used to use the car as everyday transport when working in London and commuting back to Oxford at weekends.

In the last 25 years the car has been very regularly maintained by John but has not needed any major restoration or remedial work whatsoever. John believes in constant regular maintenance and cleaning, which has meant that his Mark VII is still totally original and in perfect working order, yet has a beautiful patina of age on the seats which makes the car very appealing.

Of course, this pre-M Mark VII is a lot to handle, but John has become accustomed to the lack of power steering, the drum brakes and limited visibility, so driving habits are adjusted accordingly. The overdrive makes the car relaxing to drive on long runs.

In conclusion, although John also owns a Mark X saloon and an E-Type sports, given the choice of only one vehicle, it would have to be his Mark VII; not because it was the first car he owned but because it is so "Jaguar" and so important in the history of this great marque from Coventry.

The chrome strip following the wing line of the Mark VIII made the car seem longer and lower. Mark VIIIs and IXs had these cutaway spats. This Cornish Grey over Mist Grey car was restored by Dave and Brenda Bower.

Passenger compartment of the Mark VIII with walnut veneered picnic tables, clock, magazine rack, nylon over-rug and individually sculpted rear seat backs.

With the launch of the Mark VIII came a revised dashboard top and some amended gauges. This is an automatic-transmission model, with column-mounted selector.

Automatic-transmission cars had a bench front seat.

The 210bhp Mark VIII engine had the turquoise-painted B-type cylinder head. Power was up by 20bhp over the Mark VIIM, and 50bhp over the Mark VII.

Flush-fitted sun visors were a unique feature on Mark VII to IX models, only repeated for a short while on the very early Mark 2s. Note the sunshine roof, a standard fitment on all flagship models from 1950 to 1961.

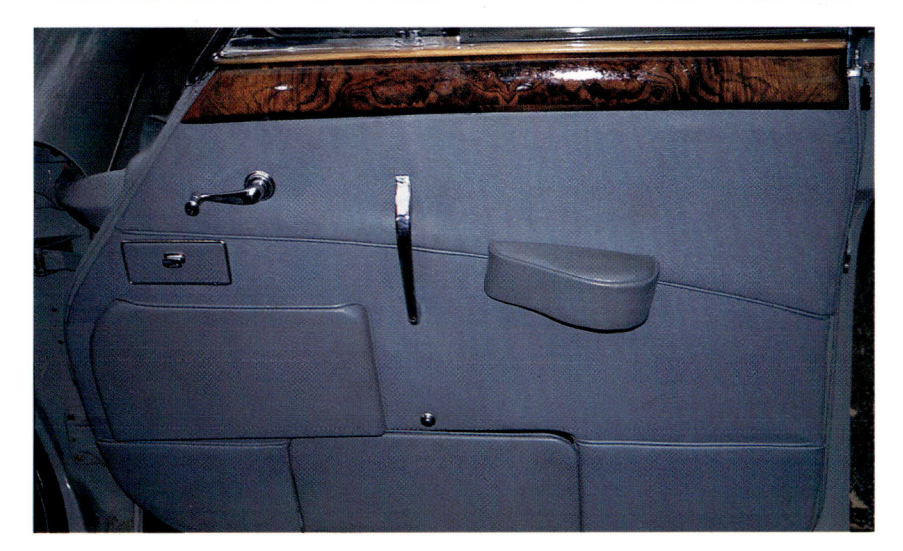

Front door trims on Mark VIIM, VIII, and IX featured shallow map pockets and these complex Wilmot-Breedon door locks incorporating trigger operation.

51

The stylish rear number plate and reversing light housing with twin boot handles used on Mark VII to IX models.

Second generation rear light, first seen on the mid-production Mark VIIs.

The dash-mounted Intermediate Speed Hold switch fitted to automatic-transmission cars.

Mark VIII
More luxury, more power

"The finest car of its class in the world" is how Jaguar described the latest addition to its model range in the opening lines of the Mark VIII colour brochure. "A luxury new model now joins the Jaguar range". "Addition" was the operative word, for the Mark VIII was conceived as an extra model to run alongside the Mark VIIM. It was much more lavish in both internal and external decoration. Greater use of brightwork was the hallmark of the Mark VIII, this meeting with North American approval, and although the interior of the Mark VIIM was considered to be in the best British tradition, the Mark VIII was positively opulent in comparison, with many detail fitments placing the car truly in the limousine class.

It was released to the public in October 1956 and was made alongside the Mark VIIM until the latter was discontinued in July 1957. The Mark VIII had a more limited production run than the Mark VIIM. Mark VIII production then overlapped the introduction of the Mark IX and petered out soon after.

The Mark VIII was, however, something rather special in its day. Although it retained the same bodyshell as the Mark VIIM, the car took on a new look with a bolder chromium-plated surround to the radiator grille, with a redesigned winged Jaguar badge on top and a brass-based plated leaping cat mascot on the centre of the bonnet. Other aspects of the Mark VIII's frontal appearance remained exactly as the Mark VIIM except for the windscreen, which was now one piece instead of the rather old fashioned split-screen. As no changes to body pressings were made, the slight "peak" in the centre of the roof remained where the split-screen centre support had been accommodated.

At the rear no changes were made at all, but from the side the Mark VIII was made to look longer and lower with the aid of a thin plated strip running the whole length of the car, following the shape of the wings across the doors and straightening out over the rear wheel spats. The spats themselves now had cutaways revealing most of the hub caps.

Most Mark VIIIs were finished in

Some styling ideas to improve the look of the Mark VII were fortunately not adopted.

The new grille surround and mascot greatly increased the presence and opulence of the big Jaguar.

The rear aspect was totally unchanged, not even featuring Mark VIII badging. The wraparound rear bumper had been introduced on the Mark VIIM.

two-tone paintwork, which helped emphasise the length of the body. The darker colour was always featured above the new waist strip, which was used as the break between the darker and lighter colours. An intriguing array of colour schemes was available, obviously chosen to encourage sales in North America, although some more muted colours were also available for the home market:

TWO-TONES

Exterior	Interior Trim
Claret over Imperial Maroon	Red or Grey
Indigo Blue over Cotswold Blue	Light Blue, Dark Blue, Grey or Red
Cornish Grey over Mist Grey	Red, Light Blue, Dark Blue, Grey
Sherwood Green over Forest Green	Grey or Suede Green

| Black over Claret | Red, Tan or Grey |
| Black over Forest Green | Red, Tan or Grey |

SOLID COLOURS

Exterior	Interior Trim
Cornish grey	Red, Light Blue, Dark Blue or Grey
British Racing Green	Tan or Suede Green
Pastel Blue	Light Blue or Dark Blue
Sherwood Green	Tan or Suede Green
Mist Grey	Red, Light Blue, Dark Blue or Grey
Indigo Blue	Light Blue or Dark Blue
Imperial Maroon	Red or Grey
Claret	Red or Grey
Cotswold Blue	Light Blue, Dark Blue or Grey
Old English White	Red or Tan
Black	Red, Tan or Grey
Pearl Grey	Red, Light Blue, Dark Blue or Grey

Secondary colour combinations known to have been supplied from the factory although never listed as standard included:

Sherwood Green over Mist Grey
Imperial Maroon over Cornish Grey
Black over Cornish Grey
Salmon over Cream

Other combinations or single colours could also be ordered specially, and it is known that Biscuit as an interior trim colour was used although not listed.

Many of the above single colour schemes were also used on other Jaguar models of the period.

Inside, the Mark VIII offered its occupants the very best of British craftsmanship, a feature that would be promoted well in the launch brochure. Firstly, the veneering on the dashboard and other woodwork had been improved, with better walnut figuring. The veneered dashboard cant-rail was redesigned without the centrally mounted ashtray, which was replaced by two ashtrays fitted flush in the front door trim panels.

The front seat height was increased to help shorter drivers see over the vast Bluemels cast steering wheel and high scuttle. The front seats were also slightly broader and now had extra Dunlopillo padding along with wool felt packing. On cars fitted with automatic transmission a single bench front seat was used, with a centre armrest. For the benefit of the driver, a much wider brake pedal rubber was fitted on automatic transmission models, enabling the use of either left or right foot for braking.

On the backs of the front seats, twin veneered fold-away picnic tables were supplied for rear seat passengers. These were beautifully made, with plated cantilever mechanisms, and veneered panelling was also featured on the upper rear edges of the front seats. On bench seat models a magazine rack in wood veneer was fitted between the picnic tables. Above the magazine rack was a Boudoir electric clock which was illuminated when the lights were switched on. The ashtrays in the rear compartment were resited in the rear door trim panels to match the front.

In order to further enhance the rear accommodation of the Mark VIII, the rear seat had been sculpted to resemble individual seats, although a centre armrest was still fitted which, when folded away, allowed space for three passengers abreast. A special nylon deep-pile over-rug dyed to match the carpet colour was fitted over the carpeting,

Two-tone (above) and rarer single colour (opposite) Mark VIIIs. Distinctive features of the Mark VIII included a one-piece windscreen, chrome swage line trim, cutaway spats, bolder grille surround, and new leaping cat bonnet mascot.

which itself was fitted on top of heavier ¼in-thick felt underlay; press-stud fitted, this over-rug could be removed and stored in the boot when not required.

Even the boot area of the Mark VIII has been upgraded by the fitment of Hardura covering not only to the floor, but also to the spare wheel, and improved board panelling to the underside of the boot lid.

Mechanically the Mark VIII was uprated with the fitting of a new B-Type cylinder head (painted turquoise between the polished cam covers), part no. SD 1051, the idea being to improve mid-range performance. This had the same diameter inlet ports as the standard head but with the bigger C-Type exhaust valves. These larger 1⅝in valves had 45-degree angled seats (instead of 30 degrees) with concave instead of convex faces to improve gas flow. A new dual exhaust system with two silencers was also used along with a revised inlet manifold with separate bolt-on water rail. Finally twin 1¾in SU HD6 carburettors were used.

These engine modifications increased power output to 210bhp at 5,500rpm and vastly improved performance. Despite a slight increase in overall weight to 36cwt the Mark VIII could boast the following acceleration times:

0-50mph 8.7 secs (9.8 secs Mark VIIM)
0-90mph 26.7 secs (33.4 secs Mark VIIM)
Standing quarter mile 18.4 secs (19.5 secs Mark VIIM)
Top speed went up to 106.5mph compared with the Mark VIIM's 104mph.

The standard compression ratio remained at 8:1 although 7:1 and 9:1 were still available as options.

The transmission remained the same as on the Mark VIIM although most Mark VIIIs were produced with the Borg Warner DG automatic transmission for the American market. The

Mark VIII brought in a Jaguar patented device which was to become the hallmark of Jaguar automatics for many years to come – the Intermediate Speed Hold. To hold the transmission in intermediate gear without resorting to full throttle kick-down Jaguar devised a neat dashboard-mounted switch operating an electric solenoid fitted to the rear oil pump incorporated in the extension case assembly at the rear of the gearbox. The car would remain in this gear until such time as the driver moved the switch to the off position.

The Mark VIIM chassis was carried over, although with slight modifications to the apertures in the cross members in order to take the new exhaust pipes. For some reason best known to Jaguar, the fuel tanks, while exactly the same as on Mark VIIMs, were painted in light buff enamel rather than grey and so had their own part numbers!

Launched at the price of £1,331 plus Purchase Tax of £666. 17s, the automatic transmission version cost a total of £1,997. 17s compared to the price of the equivalent Mark VIIM of £1,903. 17s. At just £94 more, the Mark VIII was a bargain!

Mark VIII: what the Press said

Due to the limited production and availability of Mark VIIIs only two road tests are known to have been carried out, one by *The Autocar* in October 1956 and the other by *Motor Sport* in November 1957.

The Autocar noted that the Mark VIII was still excellent value. Much of the road test was devoted to re-stating the excellence of the Jaguar suspension system and the driveability of the car, but as the Mark VIII tested was fitted with the Borg Warner automatic transmission with Jaguar's new Inter-mediate Speed Hold, the opportunity was taken to assess the new device.

The testers found the throttle return

spring on the accelerator very heavy (a common complaint with Jaguars), making it difficult to manoeuvre the car steadily in heavy traffic conditions. This also proved a hindrance when using the Kickdown facility to bring in intermediate gear for fast acceleration or hill climbing. However, with the aid of the Intermediate Speed Hold switch, which they found ideally placed for the driver's right hand, it was easily possible to select and maintain intermediate gear whenever desired.

They found the driver's vision improved by the single-piece windscreen and the increased height of the driver's seat, although vision was slightly impaired by the thick side pillars carried over from the Mark VIIM.

The wider brake pedal rubber was also welcomed, particularly for those drivers who wished to use the left foot to brake. The change in ashtray position was liked, although in the case of the driver it was all too easy to drop ash into the map pocket rather than the tray.

The new rear compartment treatment was much appreciated although some loss of foot room was noted, particularly with the front seat pushed right back; a letdown, according to *The Autocar*, in such a large, luxurious car. As far as exterior styling was concerned *The Autocar* preferred the "new look".

The road test's concluding comments should be repeated verbatim: "...it is difficult to conceive better value for money than is offered by the Mark VIII Jaguar. Certainly it has no close rivals on such a basis on this side of the Atlantic. With five years of world-wide customer approval behind it, and backed by racing experience which has proved the high quality of its power unit, this latest model is outstanding by any standards".

In an evaluation by *Motor Sport* it did not initially appear that the Jaguar was going to be liked by the Editor, more used to the sporting machinery that befits the *Motor Sport* image. However, after a few miles accustomising himself to the bulk and the company director image he appears to have found the car appealing.

Apart from a lack of vision around the screen pillars and the squeal of the tyres when cornering hard (rectified by increasing the tyre pressures) the car performed well, aided by the Intermediate Speed Hold which, as the Editor put it, "...provided a maximum of over 80mph in this gear, which disposes of long columns of perambulating tinware."

The Editor thought the Mark VIII the equal in luxury of cars twice its price, only a few intrusive rattles betraying its low price. Another slight criticism concerned the speedometer needle, which hid the total mileage reading when in the 80-90mph sector!

In conclusion, however, the Editor gave the car its due: "...this is not a car for the Toads of this world but it represents fast, safe transport for all, save morons... I really wouldn't mind a Mark VIII myself. Mon Dieu, I must be ageing."

Jaguar's rivals

By the time of the introduction of the Mark VIII Jaguar in 1956, Alvis had introduced their TC108/G 3-litre two-door coupé with bodywork by Graber. Elegantly designed and giving 104bhp, the Alvis was sportier than the Jaguar but a lot more expensive. The successful Armstrong Siddeley Sapphire 346 continued unchanged and remained a strong competitor to the Jaguar, although without the interior luxuries of the Lyons car.

By this time Rolls-Royce had introduced their new standard steel S type Bentley which, although a beautifully elegant contender for the luxury car market, was significantly more expensive than the car from Coventry. With a larger engine (4,887cc) the Bentley didn't offer any advantage in performance over the Jaguar and did not handle as well, but in size and stature the Bentley was easily a match for the Mark VIII.

Daimler offerings amounted to the One-O-Four saloon in 3½-litre 137bhp form, which arguably looked more modern than the Jaguar, being a shape developed in the 50s rather than the 40s. There was even a Ladies' Model which offered numerous standard fitments including electric windows, matching luggage, and more. The Majestic came on the scene utilising virtually the same chassis (a pre-war design) but with new and controversial bodywork. Though aimed fairly and squarely at the Jaguar market it didn't offer the same turn of speed or ride comfort.

Austin, remarkably, introduced the new Princess IV model with streamlined body by Vanden Plas. Offering a high degree of luxury and with a 4-litre 150bhp engine the Princess came close to the Jaguar's performance but lacked the presence and parentage of the Mark VIII.

Rover introduced the 105R model developed from the earlier P4 range, this time with a special Roverdrive automatic transmission with overdrive. Cheaper than the Mark VIII but smaller and certainly less powerful, the Rover still tended to cater for the more professional type of owner.

In keeping with Jaguar's move to two-tone paint finishes and more brightwork Riley tried the same treatment with the Two-Point-Six, a modernised version of the older Pathfinder model fitted with a six-cylinder BMC engine. Lower priced than the Jaguar, the 2,639cc engined Riley only produced 101bhp but could match the Jaguar's top speed.

In the North American market, competition was provided by models like the Buick Century and Roadmaster, and the Chrysler Saratoga and Imperial Crown, all of which offered more brake horse power and higher specifications including cruise control, electric seat adjustment and the like, but there was no getting away from the sheer driveability of the Jaguar and its leather-and-walnut ambience.

Production changes

Obviously, with a limited production run the Mark VIII model was not subject to as many production changes as other models. Very shortly after production commenced, the exhaust down pipes and intermediate pipes were reduced in diameter. This meant that when replacements were ordered special adaptor sleeves were needed reducing the outer diameter from 2in to 1⅞in.

After some complaints of a rubbing sound from the rear when the car was fully laden, the problem was found to be that the drain pipes leading from the sunroof were too long and fouled the rear tyres. They needed to be cut back to eliminate the problem.

In February 1957, from engine no. N.6662, the engine was fitted with drilled camshafts to reduce tappet noise when starting from cold. In May Mark VIIIs (and some Mark VIIMs identified earlier) were fitted with new rear road springs with synthetic rubber buttons between the leaves. This affected Mark VIIIs from chassis nos. 760476 rhd and 780462 lhd. Also in May, from engine no. N.7197, automatic transmission cars received a new solenoid for the Intermediate Speed Hold, part no. C.12740. The new fitting had a three-point mounting and a larger diameter plunger.

In July 1957 Jaguar introduced an economy measure into the production of Mark VIII radiator grilles and leaping cat mascots. On all cars manufactured from chassis nos. 761116 rhd and 780870 lhd they were produced in die-cast alloy instead of brass. In the same month all cars received a new, stronger wiper motor (DR 3 type) effective on Mark VIIIs from chassis nos. 760989 and 780777.

In September all Jaguars including Mark VIIIs were fitted with a smaller dynamo pulley and shorter fan belt to increase the dynamo running speed, effective from engine no. N.8974.

In January 1958, at chassis nos. 762728 rhd and 781180 lhd, a voltage regulator type RB.310 (with black metal cover and only three terminals) replaced the RB.106/1 previously used. This necessitated the fitment of an amended wiring harness.

By February 1958 Jaguar had revised their exterior paint range to cover a range of colours not specific to any particular model, except for two-toning which was always a standard option on Mark VIIIs. The range, all in synthetic enamel, was then as follows:

Exterior	Interior Trim
British Racing Green	Tan or Suede Green
Pearl Grey	Red, Light Blue, Dark Blue or Grey
Imperial Maroon	Maroon
Old English White	Red, Light Blue or Dark Blue
Indigo Blue	Light Blue, Dark Blue or Grey
Claret	Red or Maroon
Cotswold Blue	Dark Blue or Grey
Black	Red, Tan or Grey
Mist Grey	Red, Light Blue, Dark Blue or Grey
Sherwood Green	Tan or Suede Green
Carmen Red	Red
Cornish Grey	Red, Light Blue, Dark Blue or Grey

Plus a standardised range of two-tone schemes consisting of:

Claret over Imperial Maroon	Maroon or Grey
Indigo Blue over Cotswold Blue	Light Blue, Dark Blue or Grey
Cornish Grey over Mist Grey	Red, Light Blue, Dark Blue or Grey
Black over Sherwood Green	Grey, Suede Green or Tan
Black over Claret	Grey, Tan or Red

In May, due to complaints of a jerk when a closed throttle downshift between intermediate and low gear took place in the automatic transmission, a modification was carried out from engine no. N.A 1938 and subsequently advised to dealers to carry out on existing cars. This involved removing and dispensing with the relay valve spring and inserting a slug between the relay valve plunger and the cover. It had the effect of cutting off the hydraulic flow to the low band servo so that only the forward band was in operation for automatic Low in the "D" (drive) position.

In July 1958 the clear plastic manual switch and relay fitted to overdrive models was superseded by a metal switch, similar in appearance to the Intermediate Speed Hold switch on automatic transmission cars. In the same month Jaguar started to offer specially designed Reutter seats as an extra cost option.

Long past the Mark VIII's production period, in July 1960, Jaguar recommended the fitting of a hydraulic check valve in the fluid line between the brake master cylinder and servo unit which maintained a small residual pressure to reduce the tendency for air to enter the system. This followed complaints from owners of long brake pedal action resulting from small quantities of air entering the hydraulic system on Mark VIIIs and VIIs.

The Mark VIII continued in production after the introduction of the Mark IX, terminating finally in December 1959 after production of only 4644 rhd cars and 1688 lhd, making a grand total of 6332 which, for a supposedly mass produced saloon, only amounted to an average monthly production of 160 cars. This would hardly have been viable for most manufacturers but for Jaguar, with Lyons' clever policy of interchangeability of parts, the Mark VIII proved its worth. It was to soldier on into the early 1960s in the form of the Mark VIIIB, a specialist vehicle made up from Mark VIII and IX parts to be discussed in a later chapter.

The Mark VIII today

It is not easy to find a Mark VIII in good enough condition to evaluate. However, Dave Bower from Wilmslow in Cheshire, a long-time Jaguar enthusiast, came across an advertisement for a Mark VIII in the Manchester area. On inspecting the car, which had been standing unused for five to six years, Dave decided to buy it, even though it wasn't that cheap and needed an awful lot of work. Although the car required major body restoration the floor and chassis were intact.

Work on the body, all done by Dave himself, included making new bottom sections for all four doors, new inner and outer sills, rear valance, bottoms of front wings, side light pods, rear wheel-arch lips and "D" posts. Only the sills and "D" posts were available as remanufactured items, from Worcester Classic Spares; everything else needed Dave's expertise in cutting, bending and shaping. Most of the brightwork was rechromed but minor items were replaced including the swage line trims (which Dave was able to buy new from an autojumble).

Mechanically, apart from a carburettor and brake rebuild, no major work has been necessary although Dave suspects that the engine was rebuilt at some time before he purchased the car.

Inside, Dave's Mark VIII looks magnificent, with original leather in grey, a new headlining made and fitted by him and his wife, original veneered woodwork, and carpets cut to original specification. The one major problem experienced involved the sunroof drain pipes, which had caused major corrosion (a common fault on these cars).

Dave Bower has done an excellent job in restoring his Mark VIII, all on a very tight budget and without outside assistance except for chroming.

Finished in Cornish Grey over Mist Grey, his car has the Borg Warner automatic transmission with Intermediate Speed Hold device and fortunately, despite its unknown history since leaving the factory in 1958, still has all its original tools.

Dave really enjoys using the Mark VIII, which he finds a very practical classic. The automatic transmission is particularly smooth and operates without undue hesitation or snatching. The car doesn't corner as well as it might due to having tyres of the wrong specification, but their replacement is high on the list of jobs for the future. When Dave first put the car back on the road he found it "weird", with marked understeer, but once accustomed to the size and feel of the Mark VIII, he found it compared favourably with his Mark 2. The Mark VIII is remarkably easy to drive and offers a feeling of great solidity and safety.

"And now the Mark IX…"

Earls Court Motor Show 1958 saw the launch of the Mark IX, visually identical to the Mark VIII except for Mark IX script on the boot lid.

A larger engine, better brakes and power assisted steering were the upgrades that Jaguar awarded the Mark VIII to justify a change in model designation to Mark IX. These developments were to feature strongly in their launch advertising: "And now the Jaguar Mark IX… characterised by phenomenal acceleration and the ability to attain high cruising speeds… accurate finger light power assisted steering ensures completely effective control at all speeds whilst the unparalleled stopping power of the race-proved Dunlop Disc Brakes on all four wheels invests the Mark IX with the highest degree of safety."

During the 1950s Jaguar had developed an excellent reputation in racing, particularly in the Le Mans 24 hour race, with several wins for the specially designed C-Types and D-Types. Jaguar had adopted an enlarged capacity for the XK engine for this purpose – 3.8 litres (3,781cc).

Due to racing success with the larger engine, Lyons decided to upgrade the production unit to 3.8-litre capacity to offer similar advantages in performance, whilst at the same time retaining the 2.4- and 3.4-litre versions.

The new engine capacity was arrived at by increasing the bore of the Mark VIII engine from 83mm to 87mm whilst retaining the stroke of 106mm. The existing engine block could have been retained, but to avoid possible problems in service it was decided to manufacture a new block with water passages between each set of three cylinders, and for the first time on any Jaguar dry liners were fitted. Lead bronze crankshaft bearings, as fitted to some of the last Mark VIIIs, were also used.

The existing B-Type cylinder head was retained, but with a chamfer at the bottom of the combustion chamber to accommodate the increased cylinder diameter. Power output was improved by 10bhp to 220bhp at 5,500rpm, the engine also delivering 11½% more torque – 240lb/ft at 3,000rpm. 8:1 was still the normal compression ratio, although 7:1 and even 9:1 were offered as alternatives.

The same twin SU HD6 carburettors were fitted, but the auxiliary starting carburettor system was revised to give more even distribution of the rich mixture by feeding it to six points under the induction manifold. The revised, thermostatically controlled unit was far more reliable, particularly in really cold climates where Jaguar had received complaints of the ineffectiveness of the old unit. A separate fuel filter was also fitted on all Mark IXs.

The Mark VIII chassis was retained for the Mark IX but fitted with four-wheel disc brakes, a most important advance in view of the weight and performance of the car. In 1958 disc brakes were still a novelty and to have them fitted to all four wheels was something that no other competitor in the luxury car market had done.

The Dunlop system had massive 12⅛in by ½in front discs, and 12in by ⅜in rears. Along with twin wheel cylinders and hefty square (quick change) brake pads Jaguar also fitted a Lockheed 6⅞in vacuum servo unit with a storage reservoir allowing the system to hold vacuum for use if the engine stalled. The whole system was well up to handling the performance of the car.

The only other mechanical change from Mark VIII to IX came in the steering department. A few of the later Mark VIIIs had been fitted with power assisted steering, and this was now developed and featured as standard equipment on all Mark IXs. It was a necessity for the US market and made the Mark IX the first production Jaguar to be so fitted as standard.

The system was based on conventional Burman recirculating ball worm-and-nut steering, the nut being extended to form a piston working within a cast iron cylinder, pressed into the steering box, which was slightly enlarged from the earlier design. It worked well, with a self-contained Hobourn Eaton pump driven off the rear of the dynamo. Although the turning circle stayed at 36ft, the number of turns lock to lock was drastically reduced from 5 to 3½.

Externally it was impossible to tell the Mark IX from the earlier Mark VIII except for the discreet "Mark IX" badge on the boot lid to the right of the number plate mounting; all other aspects remained exactly the same, as Jaguar clearly felt that no further development of the existing bodyshell could take place (although some unwieldy attempts were made by the styling department, fortunately in drawings only!).

The interior remained unchanged

Early Mark IX stern showing the badging and the small rear lights.

The dashboard of the Mark IX. The indicators were now controlled by a column mounted stalk and there were two-speed wipers.

except that in the rear compartment of bench-seat models the veneered magazine rack was now hinged and lockable. The fitted lambswool over-rug for the rear seat passengers was also retained. Special Reutter reclining seats could be specified as an extra cost option for the front compartment.

Jaguar had received many complaints about the inefficiency of their heating systems. They tried to redress this by fitting an upgraded system for the Mark IX, with a two-speed fan operated by a single knob switch on the dashboard, pulling out for half speed, turning and pulling out again for the higher speed. A new heater with redesigned matrix was accommodated in the centre of the bulkhead directly below the scuttle ventilator intake, the vertically mounted two-speed fan directing air via the under-dashboard ducts as well as to the screen. The new system gave an improved output of 5 kilowatts (3½ on the Mark VII/VIII) with an increase from 110 to 150cfm air flow. To accommodate the new heater box centrally on the bulkhead the battery had to move, necessitating the fitment of two 6 volt batteries in series astride the steering column.

The Mark IX was released to the public in October 1958 at a basic price of £1,994 in manual transmission form, a marginal increase over the £1,885 price of the Mark VIII, although with overdrive the Mark IX would have cost £2,062 and with the Borg Warner automatic transmission the price went up to £2,162. Mark IXs were therefore the first Jaguar saloons to exceed the £2,000 price barrier.

In performance terms, whilst in theory the acceleration times should have been significantly better, in comparative road test reports the Mark IX had only a slight edge over the Mark VIII up to 60mph, but the increased torque helped the 3.8-litre car in the higher speed ranges and gave it a genuine extra 8mph on the top speed. The performance figures were:

	Mark IX	Mark VIII
0-50mph	8.5 secs	8.7 secs
0-60mph	11.3	11.6
0-100mph	34.8	35.7
Standing ¼ mile	18.1	18.4
Maximum recorded speed	114.3mph	106.5mph

It is interesting, at this stage, to make a direct comparison between the performance of the original Mark VII and its last development, the Mark IX. Although the cars had remained substantially unaltered in styling terms and indeed had weathered the many fashion changes over ten years well, Jaguar had actively sought to keep the cars up to date in performance terms:

	Mark VII	Mark IX
0-50mph	9.8 secs	8.7 secs
0-60mph	13.7	11.3
0-90mph	34.4	25.9
Standing ¼ mile	19.3	18.1
Maximum recorded speed	101mph	114.3mph

Interesting comparisons – nearly a 14% increase in top speed, although the price had also increased considerably from £1,276 to £1,994.

As there were no changes to the exterior specification of the Mark IX from the Mark VIII except for a scripted "Mk IX" on the bottom right hand side of the boot lid, the range of colour schemes remained virtually unchanged.

Exterior	Interior Trim
Pearl Grey	Red, Light Blue, Dark Blue or Grey
Imperial Maroon	Red
Cream (Old English White)	Red, Light Blue, Dark Blue
Indigo Blue	Light Blue, Dark Blue or Grey
Claret	Red
Cotswold Blue	Dark Blue or Grey
Black	Red, Tan or Grey
Mist Grey	Red, Light Blue, Dark Blue or Grey
Cornish Grey	Red, Light Blue, Dark Blue or Grey
Sherwood Green	Tan or Suede Green
British Racing Green	Tan or Suede Green

Two-Tones

Black over Claret	Grey, Red or Tan
Black over Sherwood Green	Grey, Suede Green or Tan
Cornish Grey over Mist Grey	Red, Light Blue, Dark Blue or Grey
Indigo Blue over Cotswold Blue	Light Blue, Dark Blue or Grey
Claret over Maroon	Red or Grey
Cotswold Blue over Mist Grey	Dark Blue, Light Blue or Grey

Many other options were available to special order as before, including among others Carmen Red, White, Black over Cream, Silver, Pastel Blue, and in interior trim Champagne (light tan) or Black.

What the Press said

Despite the lack of styling changes to the Mark IX the new model seemed to provoke a lot of attention from the media, probably due to the improved and innovative specification. The first magazine to test the Mark IX was *Autosport* in February 1959, with John Bolster giving his usual highly critical and colourful report.

Echoing his previous comments on the top-of-the-range Jaguar saloons, he

Manual transmission versions of the Mark IX featured individual front seats, with Reutter reclining mechanisms available at extra cost.

wrote, "I have long ago given up wondering how they make them for the money; for sheer value there is nothing to compare with them in the high performance field". Claiming the Mark IX to have the best all-round performance of any full six-seater in the world, Bolster considered the Jaguar represented a transportation investment almost beyond comparison. In comfort terms he had no complaints other than that rear passenger legroom was not excessive for a car of the size and he still complained of inadequate seat height for small drivers despite the wind-up height adjustment fitted.

Comment was passed on the bonus of power steering for town work, the steering being now sufficiently high geared to subdue the rear end when breaking away in icy conditions.

The quietness and smoothness of the car was praised although it was not as quiet as the smaller Mark I on certain road surfaces. In compensation the disc brakes were found to be beyond criticism.

The Motor carried out its road test of the Mark IX in October 1958, followed by *The Autocar* in December.

After a 1,300-mile road test covering the UK and Europe *The Motor* summed up its findings by saying, "...what other model, or make, of any nationality, or at any price, combines space for six passengers, a really large luggage locker, automatic transmission, power steering, disc brakes to all four wheels, and the ability to reach 90mph from rest in considerably less than half a minute?".

Whilst *The Motor* recorded up to 25mpg at times, over the whole trip the testers only managed to maintain around 13½mpg. Covering such a high mileage, *The Motor*'s team were able to evaluate the car better, as an everyday owner would. For example, they praised the siting of the spare wheel and jack in the boot so that if the boot were lightly packed with luggage access to the spare would be easy, without the need to remove anything else. Likewise the tool trays in the front doors were found to be extremely handy.

The Motor found the brake pedal pressure excessive for a car fitted with servo assistance. They also calculated that in theory the 0.82g stopping power of the Mark IX would be below that found on many drum braked cars but emphasised that the brakes were highly effective and did not suffer from fade at all, slowing the car without difficulty from 100mph to 30mph in less than fifteen car lengths on a dry road.

They also stressed, however, that due to canting of the discs under side loads, the first push on the brake pedal after a quick corner was found to be ineffective, requiring a further single pump to restore normal braking. This would have been due to loose wheel bearings.

The power assisted steering was enthusiastically endorsed by the testers, although the same could not be said of the excessive wind noise at high speed, a common trait on this series of Jaguar and a telling story of the lack of attention to this type of problem in the 1950s.

The Autocar road test also involved a considerable mileage, 1,426 miles, during which time they achieved an average 14.4mpg with a worst figure of 11.5mpg. They commented on the virtues of the power assisted steering for low speed motoring but were not so enthusiastic about its sensitivity during high speed cruising, particularly on fast winding roads. They did comment, however, on its better feel than the conventional American type of power assistance, and liked its strong self-centering action.

For the first time in a road test *The Autocar* passed comment on the new heating system in the Mark IX and found it more effective than on the previous models.

The Autocar concluded with the comment, "...The Jaguar Mark IX surely provides more of the requirements for rapid and effortless travel than any of its commercial competitors. The quality of its machinery and coachwork is much more than 'skin-deep'. Although this largest Jaguar is based on a design introduced back in 1950, progressive development has kept it abreast of the times; it is still a product of which this country's motor industry is proud."

The Mark IX didn't make it to the Australian market until towards the end of 1960 and it was not until 1961 that *Wheels* magazine got their hands on one for a road test. It was one of the later models, externally identifiable by larger rear lights.

Their test car, a relatively new example with 15,000 miles on the clock, seemed to be in need of a tune-up, particularly as the thermostatically controlled choke system was switching itself on too frequently. Another consequence of the state of tune of this particular car was that they couldn't obtain more than 110mph on a two-way test strip.

Acceleration, however, exceeded their expectations. The brakes also proved excellent, the tester "standing" on the brakes from maximum speed and slowing the car so rapidly and surely that he felt he could have removed his hands from the wheel.

In conclusion, *Wheels* found the Mark IX everything a luxury car should be, even in 1961 when the car was shortly to go out of production. Their only real criticism was the poor visibility, a problem that was becoming less common in cars of the early 1960s. The Mark IX was beginning to show its age.

The competition

By the time of the introduction of the Mark IX one in every three families in the United Kingdom had a car. Many of the cars that stole the limelight in 1958 and 1959 were not from the luxury sector, the Mini and the Triumph Herald being two that deserve their rightful place as landmarks in automotive design. On the Jaguar front, 1959 saw the introduction of the Mark 2 compact saloon, which "stole the show" in the performance sector, so the release of the Mark IX at the end of 1958 using exactly the same bodyshell as its predecessor didn't make much impact on the public, even if the media did take to the new car.

Other new models of 1959-60 which could be considered rivals to the Mark

Rear passenger compartment of a manual transmission Mark IX.

The Daimler Majestic Major (above) proved to be a worthy competitor for Jaguar's Mark IX (opposite). Both offered 220bhp and 120mph top speed.

IX included the new version of the Armstrong Siddeley, the Star Sapphire. Like the Jaguar, this had virtually the same bodyshell as previously, but with front-hinged front doors, cutaway spats, a new 3.9-litre six-cylinder engine, disc brakes on the front and power assisted steering. It was certainly a direct rival to the Mark IX, but was produced in even smaller numbers and sold at a higher retail price.

The Daimler Majestic could also be considered a rival to the Mark IX, with a 3.8-litre six-cylinder engine, automatic transmission and disc brakes, but it had a rather staid image, a pre-war chassis and questionable rear end styling. By 1960 the car had been substantially updated to Majestic Major specification, with a brand new 4.5-litre V8 engine developing 220bhp at 5,500rpm (exactly the same as the Jaguar), giving the Daimler a more sporting image with genuine 120mph performance and acceleration to match. With improved interior styling and a larger boot the Daimler was the equal of the Jaguar in all respects, and after Jaguar's takeover of Daimler the car was abandoned, possibly for this reason among others.

During this period Rover also introduced a brand new model, the 3 litre, with unitary body/chassis construction and modern slab-sided styling. Less powerful than the Mark IX, the car had a luxury image and was well priced within this section of the market.

Rolls-Royce and Bentley, although in a different price bracket to the Jaguar, were obvious contenders, and for 1960 introduced the re-designed Cloud II and SII with 6,230cc V8 power unit, power assisted steering as standard, and an upgraded heating and air conditioning system. This was a very powerful model bringing the staid old Silver Cloud shape into the 1960s.

By 1959 Rootes Group had introduced their redesigned Super Snipe

Bodied by Ghia in Zurich, this 1954 Mark VII now resides in a museum in France.

This famous Mark VII, registered LHP5, was prepared for racing by the factory and enjoyed considerable success in the hands of Stirling Moss and others. It now belongs to Allan Lloyd and has been comprehensively restored.

A Mark VII cornering at Silverstone in the 1970s.

The magnesium-alloy bodied race and rally Mark VII, now restored. Note the bonnet louvres, peg-drive wheels and side exhausts.

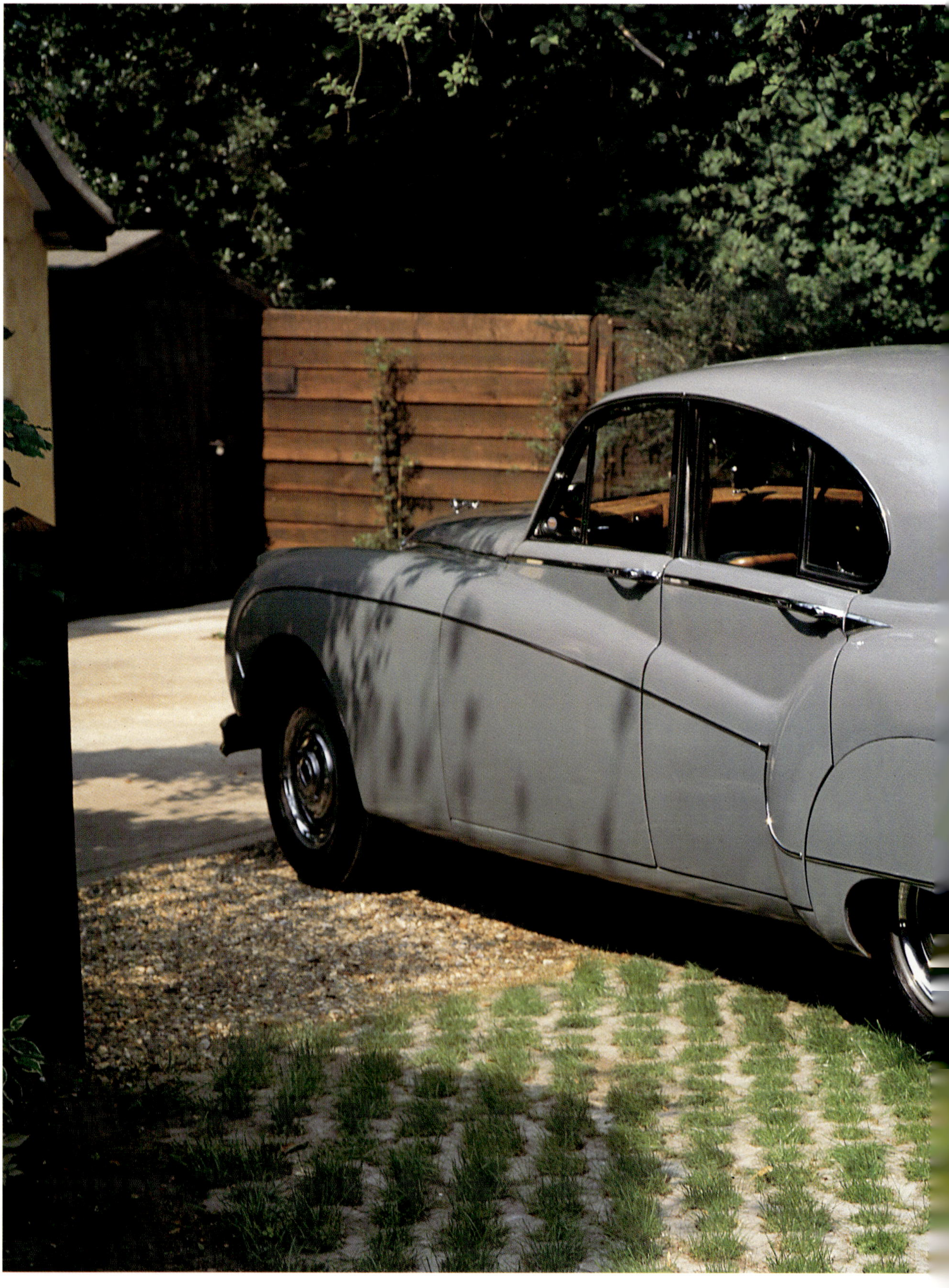

Later Mark IXs had these larger rear lamps.

The imposing front end of the Mark VIII/IX with substantial radiator grille surround.

Series II with larger 3-litre engine, front disc brakes and mid-Atlantic side stripes. Although much cheaper that the Mark IX and already well established as an official car, the Humber couldn't offer quite the presence of the Jaguar or its turn of speed.

Production changes: 1959

In January, after only about 200 cars had been produced, the Mark IX received a modification which was carried out on all Jaguars of the period: modified upper ball joints were fitted with a larger diameter ball and an increased angle of movement. At the same time an "O" ring in the power steering inner column and valve assembly was superseded by an oil excluding sleeve, retaining washer and circlip.

In April, at chassis nos. 770927 rhd and 790559 lhd, a modified banjo bolt was fitted at the top end of the steering unit to obtain greater depth of thread engagement. From engine no. NC.2198 a shielded bearing replaced the previous type fitted at the rear of the dynamo.

In May all Jaguar saloons were fitted with Mintex M33 material friction brake pads identifiable by red and white striping, replacing the Ferodo type previously used, and from chassis nos. 771237 rhd and 790713 lhd a 25 amp output dynamo and voltage regulator were fitted.

Another production change that affected all Jaguars at the time came in June when an electrically operated rev counter replaced the previous mechanically driven type. The new rev counter was energised by a small generator driven from the rear of the inlet camshaft. The cylinder head, inlet camshaft, inlet camshaft cover and gasket were modified to suit the new arrangement. On Mark IXs, this change became effective from chassis nos. 771820 rhd and 791072 lhd.

Because of complaints of a weakness in the clutch slave cylinder bracket a stronger type was fitted from chassis nos. 771827 rhd and 791081 lhd.

In October, after complaints of excessive rear end noise supposedly attributable to the rear axle, testing took place which identified that the problem could be caused by noises conducted up through the sliding roof drain tubes in the rear wheelarches. In the same month, from chassis nos. 772081 rhd and 791442 lhd, modified hydraulic dampers of the CSV type were fitted at the front, giving more consistent damping at all operating temperatures. Also in that month, from the same chassis numbers, modified rear handbrake calipers of a stronger section incorporating Mintex M34 handbrake pads were fitted. A brass retractor was also fitted to each handbrake to keep the handbrake pads clear of the disc when the handbrake was in the "off" position.

From November, all Mark IX exhaust systems had the tailpipes clipped to the silencers instead of welded.

1960

In January, commencing at engine no. NC.6785, a modified crankshaft rear cover assembly with an eccentric oil seal groove was fitted at the rear of the cylinder block. In conjunction with this a new cork seal was also fitted. January also saw the standardisation of new-style rear light lenses with separate flasher and tail light bulbs. Although the lenses were exactly the same as used on other contemporary Jaguars the plated Mazak plinths were totally different from those on XKs and Mark 2s, and were not interchangeable because of the different rear wing shaping. No exact chassis numbers are quoted by the factory for the changeover. In the same month, from chassis nos. 773392 rhd and 792257 lhd, a new type of brake master cylinder was fitted.

March saw the adoption of notched fan belts for all Jaguar engines, effective on Mark IXs from engine no. NC.6231.

In May Jaguar amended the range of colour schemes available for the Mark IX to the following:

Black
British Racing Green
Pearl Grey
Sherwood Green
Indigo Blue
Cotswold Blue
Cornish Grey
Old English White
Mist Grey
Imperial Maroon
Claret

Two-tone colour schemes for the Mark IX were now reduced to the following:

Cornish Grey over Mist Grey
Black over Claret
Black over Sherwood Green
Indigo Blue over Cotswold Blue
Claret over Imperial Maroon

Also in May, for all Jaguar engines, the top dead centre mark on the front of the crankshaft damper was now continued on the edge of the damper to facilitate tuning the engine by means of Crypton "Motormaster" testing equipment.

Not all Mark IXs were fitted from new with a handbrake/brake fluid level warning light, this only being fitted from May 1960. The light operated when the ignition was switched on and the handbrake applied. When the handbrake was released the light went off unless there was a lack of fluid in the brake fluid reservoir. It was fitted on the driver's side dash top rail to the left of the overdrive switch or Intermediate Speed Hold switch. Subsequently the switch and its wiring became available to dealers for fitment to earlier models.

November saw the transfer of the boss in the cylinder block for the fitment of a Bray electric engine heater from the left hand side to the right hand side to avoid obstruction from the exhaust pipes when fitting such a heater. This became effective on Mark IX engines from no. NC.9709.

1961

In February, due to complaints of transmission judder on take-off, Jaguar fitted a modified propshaft centre mounting assembly with larger rubbers from chassis nos. 775752 rhd and 793764 lhd. In the same month, only affecting cars destined for the USA and Canada (from chassis no. 793734) a higher output (35amp) dynamo was fitted, necessitating moving the power assisted steering hose and pump.

In March Jaguar started to issue special continental touring kits for the Mark IX called "First Aid Kits". These kits included fan belt, two 50 amp fuses, distributor contacts, condenser, rotor arm, brake master cylinder repair kit, clutch master cylinder repair kit, cylinder head gasket, inlet manifold gasket, exhaust manifold gasket, camshaft cover gaskets and oil pipe washers.

During the last few months of production all Mark IX engines were fitted with B-type cylinder heads without the machined chamfer on the combustion chamber. In June, from chassis nos. 775894 rhd and 793867 lhd, a modified type of SU fuel pump with a shorter coil housing was fitted. It is also understood that a few Mark IXs left the Jaguar factory with the Mark 2 style air cleaner and ducting assembly fitted, although it is not known how many were involved.

No other production changes were recorded and the last Mark IX left the factory in September 1961 after a total production of 10,005 vehicles, made up of 5984 right hand drive and 4021 left hand drive models.

The Mark IX Today

Although all Mark VII bodied cars have been underrated by enthusiasts of the Jaguar marque; aficionados do seem to consider the Mark IX the best model to own. Values seem to reflect this and it is interesting to consult a long-time owner of a Mark IX, Con Tennison from Norfolk.

Con is perhaps the ideal person to talk about Mark IXs as he has owned his since new back in 1959. His car is a manual transmission model with overdrive, finished in Cornish Grey with dark blue upholstery. It has now covered a total of 58,000 miles and has been Con's only car since purchase.

Originally the car was used infrequently, as Con's job took him all over the world, so the Mark IX was garaged for many months of the year. The car now covers around 1,500 miles a year as local transport during Con's active retirement; he is now over 80. Still very young at heart, he continues to "wind her up" to over 90mph on some of the excellent country roads in Norfolk.

Con carries out most of his own maintenance which has included many updates from original specification, bringing his Mark IX into the 90s. As Con says he doesn't want the "old girl" breaking down at his age!

Firstly he fitted a higher back axle ratio of 4.09:1 (instead of 4.55:1); this gives a fuel consumption of around 21mpg and makes the car a good cruiser. With the fitment of twin 2in SU carburettors acceleration is also improved dramatically, and bhp is slightly increased by the fitment of a Kenlowe electric fan. Most of the electrics are now duplicated, including

Con Tennyson with the beloved Mark IX he has owned from new.

twin transistorised ignition systems and twin coils.

Con has always found the disc-braked Mark IX perfectly adequate even with his spirited style of driving although he has converted the system to accept Silicon fluid.

Other modifications include a straight-through Servais stainless steel exhaust system; a "Hot Start" system to relieve the wear and tear on the engine through constant cold-starting; a manual over-ride to cut off the choke; and Marchal halogen headlamps with the dip bulbs (when not in the dip position) connected in series with the head bulbs to give a total of 110 watts for each lamp.

Although over the years the bodywork has suffered somewhat in the usual places (sills, front wings and door bottoms) these have all been rectified, and with regular use of Waxoyl Con is ensuring the future of his car. Interestingly, the chromework is all original and still in perfect condition without any of the usual pitting of the Mazak mouldings.

I purchased my own Mark IX back in 1982 from its third owner (another enthusiast) with a reputedly genuine 65,000 miles on the clock. Finished in British Racing Green with tan interior, this was also a manual transmission variant with overdrive. Although the car had been resprayed previously it was in basically original condition except for the fitment of Range Rover tyres which did little to improve the handling or ride and were very noisy on certain types of road surface.

At the time the Mark IX was second classic to a Mark 2 saloon, but it soon started to receive more regular attention as a good driver's car. On one occasion it was used on a foreign trip through Belgium, Germany and Austria, quite happily keeping pace with the local BMWs and Mercedes as well as the many E Type Jaguars in the convoy. Averaging around 17mpg of fuel and 750mpp of oil the Mark IX behaved impeccably, with a very good turn of speed and disc brakes to match.

To me the Mark IX, although having a very period feel of the 1940s and 1950s, is very adaptable and well capable of being used regularly even in today's heavy traffic conditions. It must surely be one of the most practical classic saloons around.

Mark VII-IX Specials and One-offs

One of the few known remaining Mark VIIIBs. This car is a 1961 example with later Mark IX rear lights.

The Mark VII bodyshell had a production run of eleven years and, although many people think it was a mass-produced car, it is amazing just how few were actually made and sold.

The "grand total" of 47,190 in eleven years is not grand by anyone's standards. To put this figure into perspective, the Rolls-Royce Silver Cloud/Bentley S Series achieved a total production of nearly 23,000, while at the other end of the scale family cars like the Ford Zephyr/Zodiac Mark 2 sold 300,000 and the Humber Super Snipe around 120,000.

Despite the small-scale production of the Mark VII-IX many memorable examples were produced, some of which still exist today. A specific model in the Mark VII/IX range as yet not discussed is the Mark VIIIB, which must now be rated as the rarest of the line and perhaps the least appreciated. The Mark VII design, because of its stature, size and competitive price, had become an ideal vehicle for official use, and just by glancing through the factory vehicle records one can identify various cars bought for this purpose: HM Queen Marie of Yugoslavia, the Argentinian Ambassador, Iranian Ambassador, Countess of Craven, Mexican Embassy, various Embassies based in Holland, the Earl of Strathmore, etc. Some of these cars were built to special requirements in either colour or interior fit-

Production numbers	Home market	Export
Mark VIIs 1950-1953 (including 1 estate and 2 convertibles)	7930	12978
Mark VIIM 1954-1957	6243	3818
Mark VIII 1957-1959	3764	2448
Mark IX 1958-1961 (including Mark VIIIB)	5362	4647

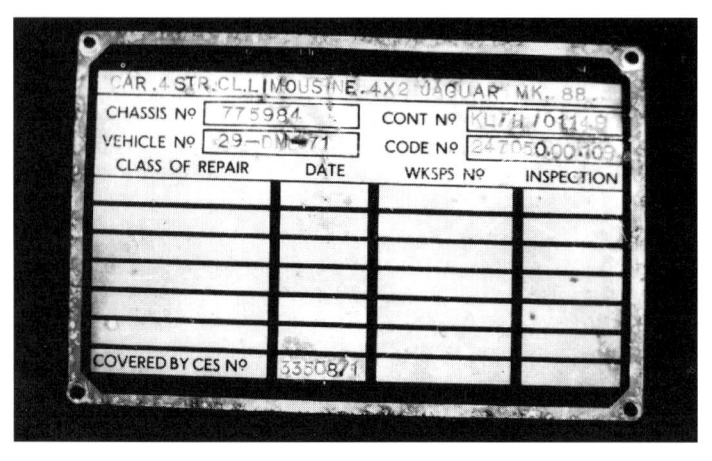

Chassis identification plate reveals the "Mark 8B" designation.

Front compartment of Mark VIIIB. Despite being a late model note the earlier horn push, incorporating indicator switch. Manual transmission Mark VIIIBs all featured a cutaway section in the bench seat.

ments. For example, during the Mark VIII period a suggestion was put forward to fit adjustable rear seating which, although never seriously taken up in production, was fitted to one car (chassis number unknown) and it certainly does not look as if that particular car exists today.

The Mark VIIIB was a development of this theme, a "special" (but nevertheless production) model ideally suited to what we can loosely call the carriage trade, although many were purchased by the War Office as chauffeured transportation. Production started towards the end of Mark VIII production and continued until the demise of the Mark IX in 1961. Mechanically the cars were based on the Mark VIII, either with manual transmission (without overdrive) or with automatic transmission. Nearly all received the 7:1 low compression engine. Power assisted steering was not fitted to any of these models.

All cars, regardless of transmission, were fitted with a front bench seat to accommodate a centre glass division with roller blind between front and rear compartments. When manual transmission was fitted the front seat squab was cut away to accommodate the gear lever, the handbrake taking the Mark VIII position under the driver's side of the dashboard. The glass division was a simple sliding affair accommodating the usual picnic tables and either a magazine rack or twin-doored compartment. The earlier Mark VIIM style full width rear seating, without the later shaped seats, was used. Most cars were also fitted with a rear window blind.

Nearly all Mark VIIIBs (to the author's knowledge) were finished in black paint, were devoid of Mark VIII/IX badging, and featured either red, tan or black interior trim. Standard options included chromium-plated standard bearers, and adaptable brackets, on the bonnet. The cars retained all the exterior trim details of the Mark VIII/IX including spot lamps, heavily chromed radiator grille, swage line chrome trims, cut-away rear spats and even the sun roof.

Whilst most of the Mark VIIIBs have now disappeared, it is refreshing to hear that one at least is in the process of being restored. Found at the Royal Ordnance Auction in Nottinghamshire (a lot of these models were sold to the military when new) and purchased by enthusiast David Marks, the car is now in the capable hands of Allan Lloyd and undergoing total restoration.

A few Mark IXs were also specially built up as limousines, retaining everything that the standard Mark IX had in both mechanical and trim elements but with the addition of the glass division, cabinets, etc.

The very first production prototype Mark VII was chassis no. 710001 which along with chassis no. 710002 was used for experimental testing and evaluation. It was later assigned for service maintenance training at the Jaguar factory before eventually being scrapped. Chassis no. 710003 was the first truly production-based model, used extensively for motor shows and featured widely in the launch programme including the Waldorf Astoria Hotel in New York. The car was finished in one of the rarer and more attractive Mark VII colours, Twilight Blue.

The next production prototype off the line was chassis no. 710004, finished in Battleship Grey and registered KRW 76. It was extensively used by the Jaguar factory for various purposes including tendering for the C-Type racing Jaguars on their many trips abroad. On one trip to Europe this car covered a total of 6,437km at 17mpg and consumed a total of 8 litres of oil. The car was laden with a host of C-Type spares, four C-Type wheels and tyres, luggage for four people, and two passengers. The springs probably took some punishment, and the tyres were recorded as having to be replaced after only 2,000 miles.

Mark VII Chassis no. 710005 was the first car exported to Hong Kong and sold there.

Chassis no. 710006 is an interesting vehicle: its body was built from light magnesium alloy and it was used for racing and rally purposes. Three of these lightweight Mark VII bodyshells were apparently built by Jaguar, although only one was actually assembled. Originally registered OVC 69, later restored and re-registered EXA 99, it also featured a race tuned engine.

The first left hand drive Mark VII was chassis no. 730001, which was exported

Limousine rear compartments: the Mark VIIIB above has a blind which can be pulled down between driver and passengers; another Mark VIIB, on the left, has a glass division and unpleated seats, but no clock or magazine rack; and the Mark IX below has ashtrays in the doors and at each end of the division, plus that enviable fitment, a cocktail cabinet.

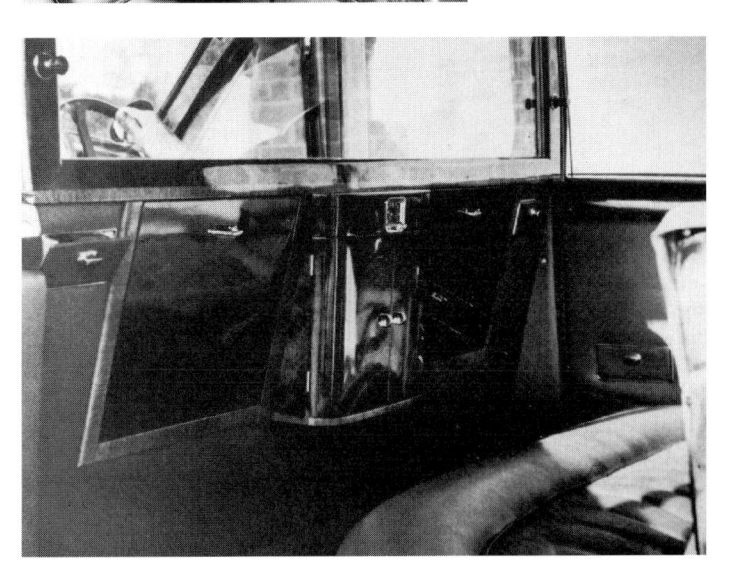

This Mark VII was converted into a rather ungainly estate in the 1970s. It was used to transport competition cars.

Her Majesty the Queen Mother's Mark VII, which was progressively upgraded to Mark IX specification.

to America and sold in California around December 1950.

Three Mark VIIs were allotted to special building, one being converted (outside the factory) into an estate car. This was a very box-like contraption and its final whereabouts are unknown. Two others were made by the factory as convertibles, one (chassis no. 750001) with a power operated hood, the mechanism coming from a contemporary American car. Neither proved successful and both were eventually dismantled.

Later in life another Mark VII was converted into an estate car by an owner who used it to transport competition cars to events in the 1970s.

William Lyons himself obviously drove a Mark VII during this period and in 1953 actually had two registered to him: LOU 268 (chassis no. 710136), finished in Battleship Grey, and ODU 849 (chassis no. 716438), an overdrive car in Black. Chassis no. 713438, registered MRW 768, was reserved for chauffeur work for the factory on official occasions and was also painted Black.

Many well known people like Lord Ogilvy, Goldie Gardner (of MG fame), Robert Morley (from the Mecca organisation) and David Nixon (the celebrated magician and comedian) owned Mark VII/IXs, but perhaps one of the most famous cars was chassis no. 727554 (a Mark VIIM) supplied to Her Majesty the Queen Mother in 1955 on loan. It was finished in Claret with grey interior and was upgraded to Mark VIII and later Mark IX specification at special request to Jaguar during the coming years. The Mark VII was one of the Queen Mother's favourite cars and had many interesting and unique features. After initial supply it was subsequently fitted with a bench front seat incorporating magazine rack, clock and picnic tables similar to later Mark VIII specification. The rear seat was covered in West of England cloth, with a bespoke centre armrest containing personal radio controls and ashtray. Companion mirrors were fitted in the rear doors, and in the driving compartment an auxiliary horn push was fitted for ease of operation by the right hand.

The car finally returned to Jaguar and was fitted with an engraved plaque on the dashboard to commemorate its period with the Queen Mother, since when it has formed part of the company's Heritage collection.

A coachbuilt Mark VII was produced with Ghia bodywork in Zurich. Originally registered in December 1954, this one-owner car, with 90,000km recorded, changed ownership to its present owner Roland Urban in the mid-70s. The car, still in fine condition, is in a private museum in Charmoy, France. Bearing chassis no. 721420-DN and engine no. D.2580, it is a handsome car.

Two Mark VIIs were bodied by Farina. One, a hardtop called the Meteor, took quite a modern approach, with two-door fixed head styling and revised radiator grille perhaps reminiscent of a contemporary Lancia. It is understood that the car now resides in Holland. The other, a drophead coupé, has not been heard of for many years.

Apart from the Queen Mother's car perhaps the best known Mark VIIM in Jaguar circles was the factory car PWK 700, finished in Suede Green. This car took eighth place in the 1955 Monte Carlo Rally, driven by Ronnie Adams, and went on to win the 1956 event out-

right, again with Adams at the wheel.

Another famous ex-racing Mark VII was LHP 5, a Dark Grey car that was factory owned and raced extensively at Silverstone and other circuits. Registered by Jaguar in July 1951, it was used by general manager Arthur Whittaker as his company car for around two years, after which it was converted for competition use.

The car was entered in various events until sold in November 1954 to Mr. D'Arcy Hughes, who ran it for many years. He finally sold LHP 5 to the well-known Jaguar enthusiast Allan Lloyd, who has now had the car totally

Rear compartment of the Queen Mother's car, with West of England cloth seat covers.

The armrest contains a panel of radio controls.

Front of the Pininfarina Mark VII known as the Meteor (above).

These three pictures show what must be the most unorthodox Mark VII ever made. Built to the specification of its owner, Dr John, this Mark VIIM now resides in the Isle of Man. The style of the rear of the car is supposed to help clear rain water. The chassis is seen below.

restored. Apparently quite basic in specification, this particular car does seem to have some added strengthening and will have had many engine modifications during its competition history. The engine was certainly rebuilt and bench tested by the Experimental Department back in April 1953, with a Weslake flowed cylinder head, high lift camshafts, 1⅞in inlet valves, twin SU H8 2in carburettors, 15½lb flywheel, and other modifications. It produced 209bhp at 5,750rpm. Other competition Mark VIIs may also have been so modified during this period.

One of the most unusual Mark VIIs now resides comfortably on the Isle of Man and still belongs to its original owner and designer, Dr John. The doctor lived in Staffordshire and was a brilliant engineer as well as a surgeon. Long involved with motor cars, he wanted to build his own to a very special and unusual specification. Contacting F.R.W. 'Lofty' England at Jaguar's factory, he was able to acquire a rolling Mark VII chassis via the then main dealers in the area, Byatt's. Onto this he grafted a highly unusual aluminium two-door sports body. Designed totally by Dr John, with modifications to the chassis, the car had many standard Mark VII trim items along with other contemporary adaptations from models such as Riley and Ford. For the radiator grille he used a cut-down version of the XK150 grille.

The bodywork, finished in bright metallic blue with tan hood and interior, was unconventional to say the least, but was effectively designed to be aerodynamic and to allow the rainwater to drain away easily without dirtying the car. With its violently upswept rear wings and bulbous sides the car caught the attention of everyone and was used regularly for work and pleasure. Dr John retired to the Isle of Man where he still maintains and uses the car regularly to this day.

One of the first experimental Mark IXs (converted from a Mark VIII) was allotted to Lockheed for brake testing, and a couple of Mark VIIIs and IXs were adapted with special bodies for funeral use by Woodall Nicholson in Yorkshire. One of these hearses was acquired by Anthony Taylor of Autotune (Aristocat replica fame) in the late 1970s and converted into a car transporter. Unfortunately the car no longer exists.

An estate version was also developed by Appleyards (Jaguar dealers in Yorkshire). It was originally commissioned by a Mr Firth of Leeds, but built in Lancashire by an unknown company. The estate, used extensively by Mr Firth in his business, was not a total success as he complained of exhaust fumes entering the car (a fault also identified when a Countryman version of the Mark 2 was made). Many years later, after the car had lain idle and rusting away, it was rescued by Tony Palmer of West Yorkshire who intends, one day, to restore it to its former glory.

This estate (registered 6826 UG) utilises most of the normal Mark IX body panels forward of the rear wheelarches although the rear door window frames have been modified. Whilst the roof line up to the B/C post is definitely Mark IX the rearward section is unique to the car. Rear wings were built specially and accommodate the fuel tanks and fuel fillers in the normal way, the conventional Mark IX rear light clusters being "added on" rather awkwardly.

The two pictures above show one of the rare Woodall-Nicholson hearse bodies built around the Mark VII bodyshell.

In desperate need of restoration but still intact is this rare Mark IX estate car, more elegant than some.

Harold Radford converted several Mark VIIIs and IXs to Countryman specification and here are just a few of the modifications carried out. Centre armrest (above) accommodated a compartment for the lady. The central division (right) could be easily removed. Front seats could be folded down (below) for weary travellers.

The two-piece tailgate was made up from a contemporary Humber estate top-hinged rear window and an adaptation of the Mark IX boot lid, bottom hinged to form the load platform.

Another interesting variant on the Mark VIII/IX theme was the Harold Radford Countryman, based around a standard factory bodyshell but with several novel features. Harold Radford (Coachbuilders) Limited of Hammersmith in London had already become well established, producing Countryman versions of Bentleys, and the Jaguar seemed a natural progression.

Various adaptations, available on either Mark VIII or IX, included a centre division converting the saloon into a limousine. The division was easily removable in a few seconds and sold for £120. Special "Countryman" front seats in the form of a divided bench with centre armrest were also available. Both seats could be reclined so that they levelled up with the rear seat to form a full-length couch. These seats were available at a price of £175, and could only be fitted to a new car.

Rear seat "Countryman" adaptations were also available, specially designed for maximum comfort. These were designed to level up with the front seats as above to form a couch or bed. When fitted, the seats and squabs folded forwards and downwards individually to form a platform for additional luggage, thus providing uninterrupted storage space from the boot forwards. There were even special strap fixings to secure luggage, and the floor of the boot compartment was carpeted to match the interior trim. The sides and boot lid were finished in rexine and moquette, all dyed to the interior trim colour. The complete "Countryman" rear seat package cost £280. Alternatively an owner could have the carpet treatment to the existing boot area of his Mark VIII/IX for just £35. The spare wheel could also be carpeted to match for £11. 11s. 0d. A hinged mirror could be supplied to fold down from the underside of the boot lid for £6. 15s. 0d.

The list of further additions to per-

Fully carpeted boot and spare wheel giving a vast load area when the rear seat was folded down.

With the rear seat folded an enormous amount of fitted luggage could be accommodated.

Boot with fitted table and picnic hamper. Note the mirror in the boot lid.

Separate sections of the rear seat could be folded.

The picnic table doubled as a stand when at the races!

sonalise your Mark VIII/IX went on. At a cost of £35 one could have a cigarette case, notebook and pencil holder, with clothes brush, comb and swivelling mirror, all finished in fine grain pigskin, in the centre armrest. Also available, for £70, was a picnic table attached to the rear of the boot which could also double as a "grandstand" as it had a top surface covered in non-slip plastic. The table would fold away under the boot floor covering. Recessed in the picnic table was a frame to accommodate a further luxury, a fibreglass washbowl with two plastic water containers – price £3.

Other picnic accessories could also be ordered, comprising a picnic case for four people holding the usual plates, cups, saucers, cutlery, Thermos flask, milk/sugar containers and sandwich boxes. The case was fitted with foam rubber to eliminate rattles and cost £30. Also available were a coffee percolator and butane heater for £13. 10s. 0d., an ice Thermos for £6. 10. 0d., and a spirit case fitted with three leather-covered flasks and six glasses for £47. 10s. 0d. All the above accessories could be stowed in the boot without encroaching on the normal luggage space.

The extras didn't end there, and returning to the inside of the car you could request rubber overmats for the front at £15. 10s. 0d., rear at £12. 10s. 0d. and even for the boot floor at £15, plus "Countryman" folding chairs at £5. Other options were a combined shooting stick/golf umbrella in a sheath that could be fitted to the rear seat heelboard for £15. 10s. 0d., a full length fishing rod container fitting beneath the doors at £30 and expanding cloth pouches between the sun visors to carry glasses, cigarettes or other small articles for £3. 10s. 0d. Lastly, a Rotaflare swivelling lamp adjustable from a control handle on the dashboard cost £14. 10s. 0d.

Harold Radford were so confident that the Countryman would be a success that they actually produced a brochure depicting a "fully loaded" Mark VIII which could be provided by many of the Jaguar dealerships. It is not known how many conversions were actually carried out to the full specification but it is believed that some 30-50 cars had some form of adaptation.

During the 1970s and early 80s several Mark VII/IXs were converted to replicas of 1930s tourers by Bill Beaman, these being known as Beaman Specials. Apparently in the beginning Bill Beaman, who had always had a soft spot for the big Jaguars and also liked convertibles, took a rusted body off a Mark IX that he had bought for just £10. After deciding that the chassis was too good and strong to scrap he retained the bulkhead and windscreen, reducing them in width and refitting them to the chassis with a shortened version of the Mark IX floorpan. The Mark IX roof panel was cut off and fitted to the main body to form a stylish rear end and, with the addition of cutaway doors and cycle wings, a 1930s-style sports tourer was created. So successful was the outcome that Bill Beaman started to produce other examples to order, under the company name Beaman Limited, in Braunton, North Devon.

Upgrades could be included in the specification including transistorised ignition, stainless steel exhaust system, roll-over bar and radial tyres. Standard specification included P100 headlights and twin rear-mounted spare wheels. The Mark IX radiator grille was used along with the original dashboard, steering wheel and seats. It is thought that around six of these cars were actually made and, in their heyday in the mid 1970s, the price quoted was around £2,200.

Another unusual adaptation of the Mark IX came about when an unknown owner had his converted to a fastback style. Having changed hands several times, the car is now believed to be in the London area undergoing restoration.

I am sure that many other Mark VII to IXs were modified over the years as the value of the cars declined. One other

Mark VII-IX Specials and One-offs

The Beaman Special is clearly a Jaguar under the skin.

This most unusual fastback Mark IX is of unknown origin but is thought to reside in London today.

example of this was the "Red Baron" built by Terry Wright as an alternative to the Bentley specials that were then being built. Using a Mark VIII as the donor, Terry went one step further by constructing his own tubular steel body frame to fit the Mark VIII chassis. Onto this he grafted a superb Invicta style touring body and was so pleased with the result that he was going to undertake the building of another similar car.

It is only possible to list details of the known examples of unusual Mark VII to IXs but there must be many more. Sir William drove Mark IXs throughout their production, they were used as course cars at Le Mans and other circuits, and many well-known celebrities both within and outside the motor industry enjoyed these fine cars.

A NEW GENERATION: THE MARK X

Styling development of the Mark X under the watchful eye of Sir William Lyons himself.

"New Grace... New Space... New Pace – A special kind of motoring which no other car in the world can offer."

Thus were the opening comments in the lavish twelve-page colour brochure for the new Mark X saloon upon its launch in 1961. It is worth repeating Jaguar's opening paragraph describing the range of Jaguar's flagship luxury saloons:

"The Jaguar Mark X, although an entirely new car in construction, design and appearance, stems from a long and illustrious line of outstanding models which have been identified during the past decade by the symbols Mark VII, Mark VIII and Mark IX. All have been highly successful in their own right and have formed important links in a chain of developments culminating in the creation of the finest car yet to be produced in the Jaguar big saloon tradition – the Jaguar Mark X."

If the Mark IX could be described as having presence, prestige and stature,

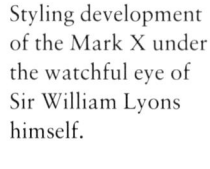

From Mark VII to Mark X. Not so much a transition as a complete change.

The Mark X retained some echoes of the Mark VII-IX roofline. It was an extraordinarily sleek car.

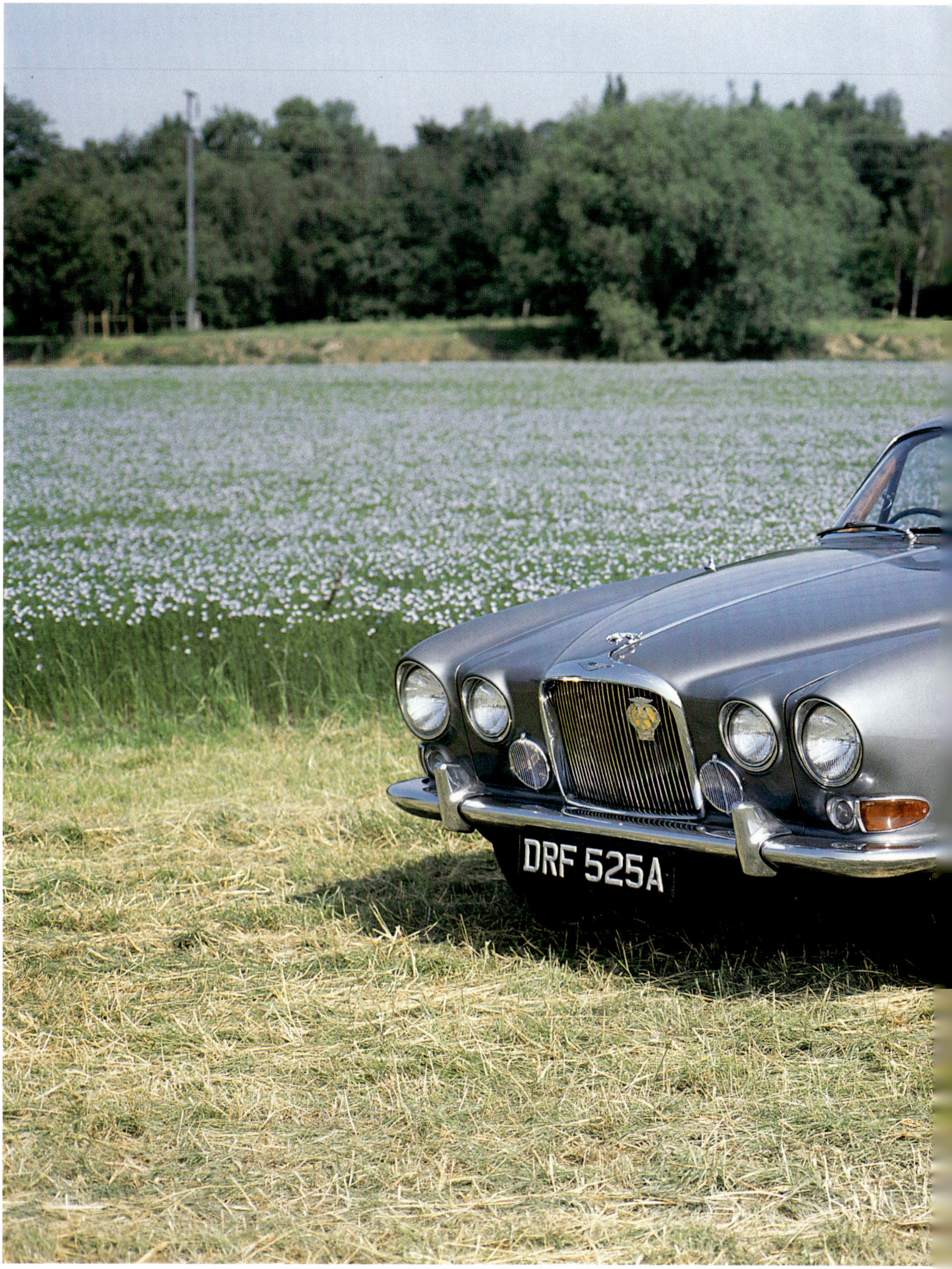
DRF 525A

The Mark X front end was completely new, but recognisably Jaguar. The bonnet was front hinged at bumper level, and incorporated the inner headlamps.

What initially seems a very spacious and comfortable front compartment suffered from not-too-easy access due to the high sills, and the wide, flat seats did not support occupants well.

The rear compartment was particularly spacious, and the seat spectacularly wide.

MARK X

Two different approaches to front end styling; fortunately neither saw the light of day.

The same car, when viewed from the side, looks pretty close to the final design. The "stylist" in the background seems as though he is just about to fit another type of front wing (more like the end result).

then the Mark X could surely be termed as lavish by any standards. The Mark X was entirely new and was not only bigger than its predecessor, but faster, more economical, with roomier accommodation, greater luggage space, more creature comforts and, because of technical advances, it even handled better.

The Mark X boasted not only an entirely new body style but monocoque construction and all-independent suspension, making it one of the most technically advanced cars of the period.

1961 was a significant year for Jaguar and with the launch of the E-Type sports car at the Geneva show, followed by the

Shaping up the all-in-one bonnet for the Mark X...

...and the finished item, now with twin horn grilles.

launch of the Mark X saloon at the Earls Court Motor Show, the company was riding the crest of a wave. A very sophisticated sports car, ahead of its time in design and economically priced, was accompanied in the same year by a most advanced saloon.

Let's take a close look at the Mark X, starting with the exterior of the car. Yes, the Mark X was bigger. Five and a half inches longer than the Mark IX at 16ft 10in, it looked longer because of its low build of only 4ft 6¾in, a considerable 8½in lower than its predecessor. The Mark X was 3.3in wider at 6ft 4in, making it the widest British production car ever made (only toppled from this position by Jaguar's own XJ220). The massive and well-engineered shell was manufactured for Jaguar by Pressed Steel in Swindon.

From many angles the Mark X could be considered well proportioned but from certain viewpoints it was decidedly bulbous, an impression heightened by the pronounced slope of the side windows in towards the roof and the "puffy" body sides. An interesting concept was the siting of the floor pan below the sills, allowing a significant reduction in the overall height of the body yet giving plenty of headroom for the occupants. Nevertheless there was no doubting that this was a Jaguar, even if you took away the traditional radiator grille, badges and mascot.

The main stiffness of the monocoque Mark X shell came from two large fabricated box-section sills bracing the scuttle and rear bulkhead assemblies. Localised box sections at the front transmitted the loads from the sills to a box section crossmember forming the scuttle, with a similar crossmember at the rear incorporating the rear seat pan structure. Stiffened inner wings at front and rear with extra members backing against the scuttle and rear seat pan added rigidity to the whole assembly.

This picture shows how the floor drops below the level of the sill to improve headroom.

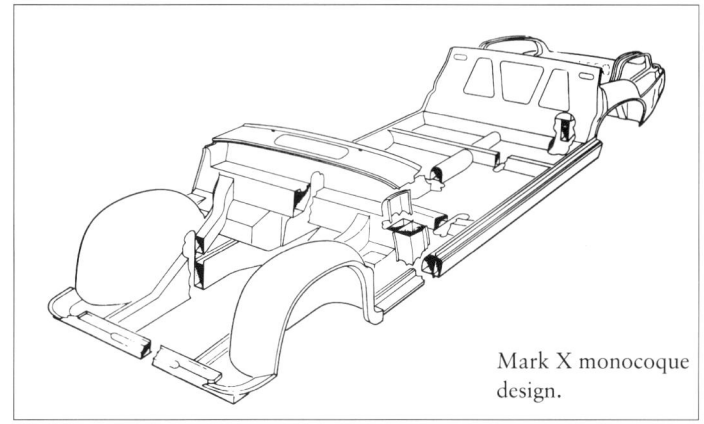

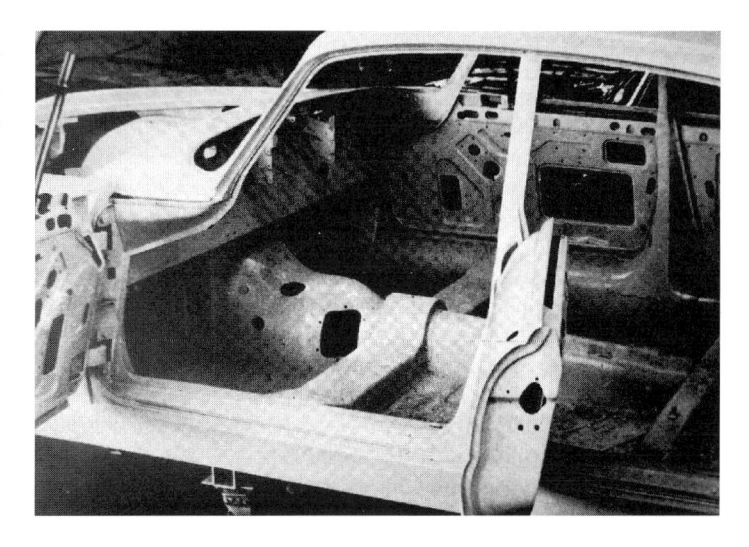

Mark X monocoque design.

Bare bodyshell showing the size of the sills and cross members.

Wraparound indicator lights for the front of Mark X/420Gs, also used on S-type and 420 models.

Prototype Mark X on test, somewhat incognito!

Torsional and beam stiffness of the shell was excellent: the inner and outer panels formed part of the load carrying structure, leaving the screen pillars and roof irrelevant to the overall strength of the body.

The rear panels of the car were joined together by welded outward turned flanges, subsequently hidden by the rear bumpers. In essence the Mark X bodyshell was relatively light yet rigid and exceedingly strong, contributing to the roadholding characteristics of the car.

Styling-wise the Mark X was to set a new trend for Jaguar. The well-known radiator grille was retained, although considerably reduced in height and raked forward. Twin chromium-plated horn grilles flanking the grille were carried over from the Mark IX, and new front side lights/indicators wrapped round so that they could be seen from the side of the car. Four-headlight treatment was used, the lights (5¾in inner and 7in outer in diameter) being slightly inset. The chromium-plated bumper bar with over-riders was of an entirely new design, much slimmer in appearance, again starting a trend that would be continued into many other Jaguar models through the 1960s. On the bonnet could be found a plated centre strip (à la Mark IX) and the traditional Jaguar leaping cat, now much smaller to suit the lower aspect of the car.

The very wide bonnet was front-hinged, giving improved access to the engine compartment. The bonnet incorporated the radiator grille and inside pair of headlights, so if spot and fog lamps were required they had to be fitted below the bumper, and a special side-of-grille mounted badge bar was available as an optional extra. Two spring catches held the bonnet shut, one either side of the bulkhead, and to ensure a first time closure of the bonnet without twisting such a large expanse of panel it was advised to close the bonnet from the front of the car with a firm hand on the centre.

The sides carried very little adornment. There was a simple swage line along the waistline of the car, new plated door handles, plated window frames and rain gutters, as before, and exactly the same hub cap and Rimbellisher treatment as on Mark IXs. The wheels were now 14in, contributing to the Mark X's low-slung lines and at the same time improving rear seat accommodation. The Mark X incidentally was the first Jaguar saloon since the war not to have rear wheel spats.

Exceptionally large and heavy doors were used on the Mark X, 40½in long at the front and 31½in long at the rear, opening on mammoth box-section hinges with helical springs inside the pillars to help occupants open the doors from the inside.

From the rear the Mark X looked massive, with an extra-large sloping boot lid featuring a substantial rear number plate housing in plated Mazak. New style lighting in chromium surrounds, new slim bumpers with wraparound almost to the rear wheelarch and new scripted Jaguar and Mark X badging were used. The under-bumper valance was more visible than on previous models and the twin plated exhaust pipes now exited just inboard of the rear over-riders. A very large rear window gave excellent rearward vision. Twin petrol fillers were accessible via flush-fitted painted lids, both lockable, on top of each rear wing.

The whole structure was immensely strong, with a recorded torsional strength of 8000lb/ft per degree, measured between the front and rear wheel planes. It was so strong, in fact, that Jaguar apparently considered producing the Mark X with pillarless construction at one time. Later in the car's life, the inbuilt strength of the shell encouraged some people to produce convertible versions.

Mechanically the Mark X relied on the tried and tested XK six-cylinder power unit. The particular variation fitted to the Mark X was based on the E-Type 3.8-litre block with straight port

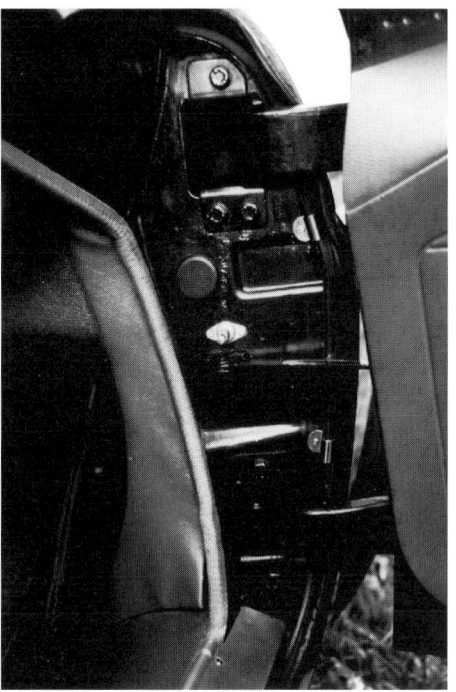

Taken at MIRA test track, this is probably a V12-engined Mark X.

The massive hinges required to support the hefty Mark X doors.

The Mark X's rear number plate mounting incorporated the reversing lights.

The rear lights on the Mark X and 420G blended smoothly into the bodywork.

The finished product in all its splendour, the 3.8-litre Mark X saloon.

"Gold Top" cylinder head (as originally seen on the XK150 "S" sports car), fed by three SU HD8 2in carburettors. The unit developed 265bhp on a 9:1 compression ratio (over 45bhp more than the Mark IX) or 250bhp with an 8:1 compression ratio engine.

The Weslake-designed straight-port head confined turbulence to a smaller area of the combustion chamber, therefore improving efficiency. There was a new inlet manifold with three separate aluminium castings (one for each of the three SU carburettors) with water jackets allowing heated water to circulate during the warm-up period. The engine used an electric choke system for cold starting. Air was taken to the carburettors from an AC Delco cleaner fitted into the right hand wing via a massive ducting pipe to a polished aluminium chamber alongside the carburettors. A cross-flow radiator was fitted, with 12-bladed 16in cooling fan.

The crankshaft ran in Vandervell VP2 lead bronze main and big end bearings, again first seen on the XK150 "S" engines, and a cast alloy sump was fitted, as on other XK power units of the

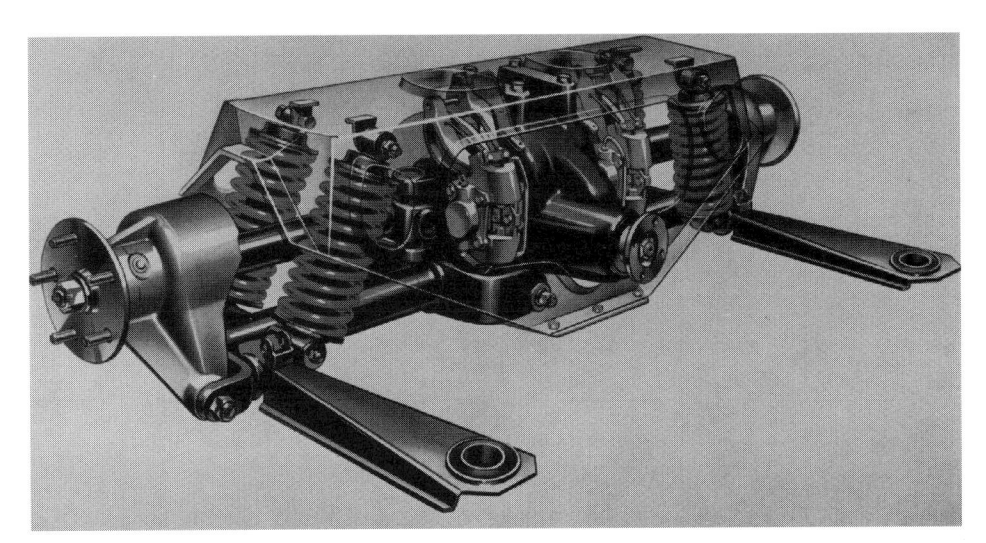

The sophisticated rear suspension subframe of the Mark X.

time, although of a different shape, a necessity due to the Mark X engine being mounted 2in further forward. The engine was also mounted 2in lower than in the Mark IX.

Not originally intended as a manual transmission model, the Mark X did become available with a choice of four-speed manual Moss gearbox, with synchromesh on 2nd, 3rd and 4th gear, or as a manual with Laycock de Normanville overdrive on top gear, in both cases with a Borg and Beck 10in hydraulic clutch with Dunlop master cylinder. Alternatively there was the Borg Warner Model DG three-speed automatic transmission as used in the Mark IX, although now with a neater column-mounted quadrant.

Drive was via a two-piece propshaft to a Salisbury HU limited slip Powr-Lok differential with 3.54:1 rear axle ratio on automatic and manual transmission models (3.77:1 for manual/ overdrive). The differential casing was solidly mounted in a 16 gauge steel pressing in the form of a cage which also held the complete rear suspension, drive shafts and brakes. This type of subframe was first seen on the E-Type a few months earlier and was used in a similar form on all subsequent Jaguars. Inboard 10in disc brakes were fitted on the first twenty to thirty cars, with 1⅝in then 1¹¹⁄₁₆in cylinders, and separate self-adjusting handbrake calipers for each disc. The entire subframe cage was easily detachable as a complete unit. It was attached to the body by two Metalastic mountings at each end, avoiding any metal-to-metal contact with the shell and thus eradicating much of the usual rear end noise.

The rear suspension was of double wishbone layout, with the top wishbones replaced by the drive shafts connected via universal joints. This system avoided adverse camber angles. Specially designed tubular forged lower arms with needle roller bearings held the cast aluminium hub carriers. Trailing radius arms located the outer ends of the lower arms and were attached to the floor via rubber mountings. There were twin shock absorbers each side surrounded by coil springs, all of which saved valuable space.

Despite its bulk the Mark X saloon sat on smaller 14in × 5½J steel wheels taking 750 × 14 Dunlop RS tyres.

The front suspension used top and bottom forged wishbones with coil springs, operating in a similar way to the Mark 2 saloons except that the wishbones were carried on one transverse beam and not a subframe. This beam was 2in deep and 1½in wide with V shaped mountings. Steering was by

Alford and Adler, and there were Girling gas cell telescopic dampers and a 0.875in diameter anti-roll bar.

Disc brakes were fitted all round, as already mentioned, with 10¾in discs outboard at the front (smaller than other contemporary Jaguar saloons). The first 1,500 cars had 2⅛in caliper pistons, later cars 2¼in pistons, with the Dunlop quick-change pad system.

The American designed Kelsey-Hayes brake servo system was used on the Mark X, made under license by Dunlop and seen on the E-type earlier in 1961. It was also fitted to the Lagonda Rapide saloon of the same period. It used a bellows system exerting mechanical instead of hydraulic pressure on the master cylinder. Basically of very simple design, the Kelsey-Hayes system was also easy to maintain and could be removed without bleeding the whole brake system. Twin master cylinders, split front and rear, gave some security against brake failure.

Steering was largely carried over from the Mark IX, with Burman power assistance via a Hobourn-Eaton pump driven off the dynamo at over one and a half times engine speed. The power steering fluid was fed from a three-pint metal reservoir and the actual hydraulic assistance equipment was integral with the steering box unit, this being mounted on the cross beam carrying the front suspension assembly. The steering column was of the divided type with universal coupling. The system gave 4½ turns from lock to lock.

Set on the same wheelbase as the Mark IX, the Mark X weighed in at 37cwt compared with the 35½cwt of the earlier model.

Mark Xs were built on the production line alongside other models with similar body treatment in the form of a six-stage phosphate process, then dipping in a corrosion resistant primer, and heating to 320 degrees centigrade, after which a bitumastic sealing compound was sprayed on the underside.

Inside, the Mark X justly earned its title as flagship of the marque with acres of veneer, leather and Wilton carpeting. The veneer featured on the dashboard, centre console, door cappings and picnic tables and was of a much lighter grain of burr walnut than Jaguar had used before. The door cappings were of similar style to those of other Jaguars.

The dashboard was similar to that of the Mark 2 saloons, with a full-width cant rail across the top and the same layout of instruments: large rev counter and 140mph speedometer directly in front of the driver, with a switch for Intermediate Speed Hold, where fitted. There was also a low brake level/handbrake on red warning light, and a neat glove box with subdued lighting and chromium plated handle on a veneered panel in front of the passenger. The fold-down centre section contained the four auxiliary instruments, from left to right ammeter, fuel gauge (registering one tank at a time via a changeover switch below), oil pressure gauge and water temperature gauge. These gauges were split into pairs by the fitment of the usual Jaguar lighting switch. Underneath were the familiar Jaguar toggle switches operating, from left to right, interior lighting (two position, first map light, second courtesy lights), panel lights (two brightness positions), heater fans (two speeds), fuel tank selector switch, wipers (two speeds) and electrically operated screen washers. Between each bank of three toggle switches could be found a cigar lighter flanked by the key ignition switch and the traditional Jaguar starter button. Below all this was an illuminated legend panel. The whole centre section of the dashboard was finished in veneer with black Rexine backing to the toggle switches. This was a move away from the metal panel fitted to Mark 2s and added to the car's air of luxury. A further little touch was the curvature of the dashboard at either end to reach the front door A posts.

Another first for a Jaguar saloon was the fitment of a full width parcel shelf under the dashboard covered in black flock and edged in vinyl to match the interior trim colour of the car. The trim continued down to form a curved centre console running back between the front seats and accommodating another veneered panel for ashtray, heater controls and a central radio opening (covered by a grille when a radio was not fitted).

There were three black pushbutton heater switches on the console. One controlled the water valve (marked "heat"), another the scuttle air intake (marked "air"), and the third cancelled the other buttons (marked "off"). For ease of operation the buttons operated on vacuum taken from the inlet manifold, aided by servo units.

The heater unit specially designed for the Mark X was derived from a Marston Excelsior heating system, using a large matrix element with the air circulated by two independent fans (one on each side of the scuttle) via pipes to manually adjusted flaps either side of the front console (allowing air to be directed to the floor or to upper body level), to twin ducts leading to the screen for demisting, and through pipes also taking air through the console to a grille with on/off flap at the back of the console for rear seat passengers. The heating system was very effective by previous Jaguar standards and had an efficient facility for ensuring air was cut off from the matrix when cooling was required.

The console, as previously mentioned, was finished in vinyl to match the other trim of the car and on either side could be found circular chromium plated grilles, the driver's side accommodating the radio speaker, the passenger's side acting as a dummy. Depending on the transmission type fitted, the centre console would also accommodate a leather trimmed gaiter for the gear lever. Further back was (where fitted) a veneered control panel for the operation of all four electric windows via small chromium plated flip switches (push forward to raise, rearward to lower) with suitable legends.

Automatic transmission cars had a steering column mounted selector stalk on the right of the wheel with illuminated legend and pointer window on top of the column, as first seen on the Mark 2 saloon in 1959. On the left (on right hand drive cars) would be found the indicator stalk, of matching style, which also acted as headlamp flasher (by pulling towards the driver). Steering column reach was adjustable via a knurled plastic ring allowing up to 4in movement to improve driving position. The 17in steering wheel with half horn ring was of a new design, accommodating a centre push button horn with Jaguar head motif, which would later be standardised for Mark 2s and all other Jaguar saloons in the mid-60s.

All seats in Mark Xs had best quality Vaumol leather covered facings with side panels finished in vinyl, and had deep Dunlopillo cushions. For the first time on a Jaguar, the front seats (always individual) had reclining mechanisms as standard equipment, operated by black pull levers beside the centre console. The seats had neat centre armrests, one each for driver and front seat passenger. These provided lateral support as the large flat surfaces of the front seats did not really help to keep driver and passenger on their own sides of the car during spirited driving.

The backs of the front seats were fitted with beautifully finished walnut veneered picnic tables with improved

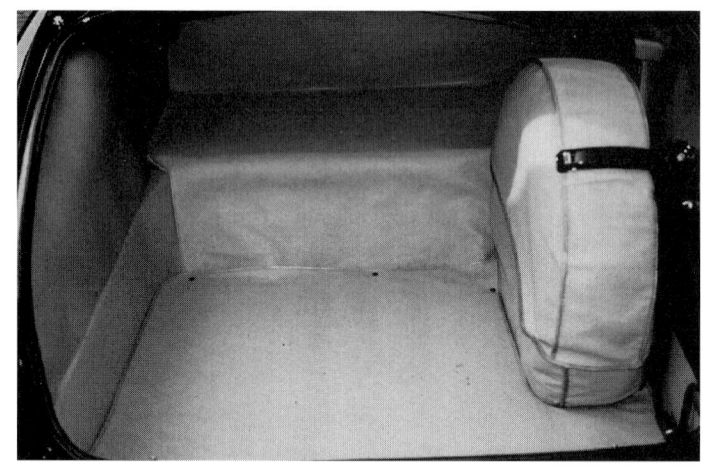

From the top: the Mark X's door trims were of relatively simple design; the vast boot was fully trimmed; and the tool kit now lived in the boot instead of in the front doors.

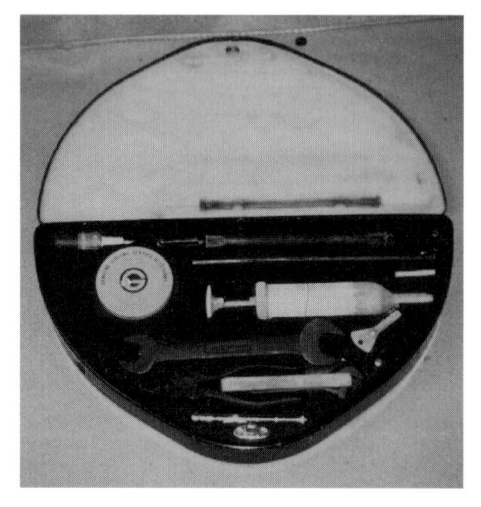

mechanisms and full width vanity mirrors when the tables were pulled open. Above the picnic tables were veneered panels with ashtrays in their centres.

Rear seat passengers were exceptionally well cosseted in a massive full-width bench rear seat trimmed in leather, with fold-down centre armrest, larger than average foot wells, not one but three roof mounted courtesy lights, and hand straps above the rear doors. Even the rear quarterlights had beautifully designed opening levers just as on the previous model. The only criticism that could be levelled at the cavernous interior of the Mark X was that getting in or out of the car was difficult to accomplish gracefully. Due to the very low floor and the very high sill, one slid into the seats in a similar fashion to a sports car, further hampered by the extremely heavy doors.

Carpets were of course of best quality Wilton, with a bound-in heelmat for the driver. Very adequate soundproofing was provided under the carpet, and Jaguar took the decision to fit an entirely new style of headlining using a sponge rubber backed material held by adhesive to a sound deadening panel that sprung and clipped into position. It was easy to fit and less costly than the type used on other Jaguar saloons.

Door panels were again finished in Rexine, with large conventional handles and window winders (except when electric windows were fitted, in which case the rear door panels featured chromium-plated flip switches) and each door had its own well-proportioned armrest.

The rear parcel shelf was also trimmed in vinyl to match the other trim, and usually accommodated to one side a further radio speaker with grille, for which a balance switch was fitted under the dash panel.

Opening the large, counterbalanced boot lid, which had tan coloured boarding and an interior light on the inside, revealed an enormous boot, though it was actually slightly smaller than the Mark IX's. The floor was covered in Hardura and the spare wheel, also covered, was accommodated upright on the right. There was a comprehensive boxed tool kit containing an impressive array of fitted tools in keeping with regular Jaguar practice. The boot had a total capacity of 27 cu.ft. Hidden behind fitted hardboard trim panels in the rear wings were twin fuel tanks with a total capacity of 20 gallons, each tank containing its own submerged Lucas electric fuel pump working on a recirculating principle and pressurising the fuel lines to reduce vapour lock.

The Mark X was released to the public at the motor show in October 1961 and directly replaced the Mark IX. Its price was £1,673 plus purchase tax of £628. 7s. 9d., and the total price of the new car in manual/overdrive form came to £2,301. 7s. 9d. Extras initially available included electric windows on all doors for £66 and Borg Warner automatic transmission for £154. Other extras also available to order included underslung rectangular matching spot/fog lamps, Desmo badge bar, Witter tow bar equipment, Sundym tinted glass for export markets, Smiths Radiomobile radio, electric or manual remote controlled radio aerials and special paint finishes. Two-tone paint schemes to customer specification could also be had.

With 3,781cc packing a claimed 265bhp at 5,500rpm, albeit in a massive 37cwt bodyshell, the Mark X was no sluggard. Performance figures were:

0-50mph 9.1secs (automatic), 8.4secs (manual)
0-60mph 10.8secs (automatic), 9.1secs (manual)
0-100mph 33.3secs (automatic), 32.9secs (manual)
Standing quarter mile 18.5secs
Maximum speed 120mph

This was quick motoring, but unfortunately slightly down in the mid-range on the previous model despite the more aerodynamic shape of the Mark X, although it made up for this on the 0-100mph time, being over 1 second faster than the Mark IX (all these figures are taken from contemporary road tests where obviously weather conditions and individual vehicles could have influenced figures).

The initial list of colour schemes for the Mark X was comprehensive, with a standard choice of fourteen exterior finishes and eight interior trim colours. Possibly excluding Carmen Red, all the colours were subtle and pleasing to the eye, not always the case with luxury cars of the 1960s.

The standard choice was:

Coachwork	Interior trim
Cream (Old English White)	Red, Light Blue, Dark Blue
Pearl Grey	Red, Light Blue, Dark Blue, Grey
Mist Grey	Red, Light Blue, Dark Blue, Grey

Earls Court Motor Show in 1961, and the crowds couldn't get near the Jaguar Mark X.

Cotswold Blue	Light Blue, Dark Blue, Grey
Sherwood Green	Tan, Suede Green
Black (Embassy)	Red, Grey, Tan
Carmen Red	Red, Beige
Opalescent Gunmetal Grey	Red, Light Blue, Dark Blue, Beige, Grey
Opalescent Silver Grey	Red, Light Blue, Dark Blue, Grey
Opalescent Silver Blue	Dark Blue, Grey
Opalescent Blue	Light Blue, Grey, Red
Opalescent Bronze	Suede Green, Tan, Red, Beige
Opalescent Dark Green	Suede Green, Tan, Beige
Opalescent Maroon	Maroon, Beige

The opalescent (metallic) colours were a new concept, first used on the Mark 2s and subsequently made available for all Jaguars of the era. They were well thought out colour schemes but the finish lacked longevity, dulling easily and awkward to match if a partial repaint was needed.

In real terms very few cars were ever produced in Carmen Red or Cotswold Blue (except for export) and the most common colours were Opalescent Silver and Gunmetal Grey, Opalescent Dark Green, Opalescent Silver Blue and Opalescent Maroon. The most common interior trim colours were grey, tan and red.

What the Press said

It was in October 1961 that *The Autocar* first reported on Jaguar's new flagship saloon. Its heading in the issue dated 13th October was, "Finest Jaguar Yet – New Mark X", and whilst this was an impression based on Jaguar's press release pack rather than a comparative road test there was no doubt that *The Autocar* was highly enthusiastic about the new model.

Press demonstrators were hard to get hold of, so the first road test of the Mark X did not come from one of the better known British motoring journals but from *Motoring Life* magazine late in 1961, its road test report entitled "Hep Cat". The opening statement – that the Mark X shouldn't do what it did as it was immensely heavy and massively large when compared to other able luxury cars like the Rover 3 litre, Mercedes 300SE or Ford Fairlane, and that the passenger accommodation was relatively poor – initially gave the impression that the testers didn't like it. However, the writer emphasised that the Jaguar was the fastest, best performing, most modest and least tuned family saloon in the whole world. He continued with a comparison to a Ferrari Superamerica which, although it attained a speed in excess of 130mph and cost over £7,000, wouldn't carry four and didn't have four doors, a golfer's luggage boot or walnut picnic trays.

Motoring Life's closing paragraph said it all: "Is it the best in the world? We believe that anyone who cares to compare the facts and the figures, and then relate them to the price, will find it hard to avoid the conclusion that among the large affairs, the Jaguar is the best value for money. And that, equally conclusively, means the best car."

Next on the scene came *Sports Car Graphics* in the USA during January 1962. Although covering most technical aspects of the new car, very little was said in detail from a driver's point of view. However, the magazine did comment that the body was tastefully well balanced and that the Mark X was an impressive, successful and badly needed modernisation of the previous model. The writer also expected a huge waiting list to buy the new car – a situation that disappointingly was not to follow!

In May 1962 *Car and Driver* came out with a "Road Research Report" on the Mark X, claiming it to be an outstanding car offering space, comfort and high performance with steering and brakes of the highest order. Jaguar was by this time a household name in North America and *Car and Driver* concluded that the Mark X held promise of even greater success for the company.

A more detailed American evaluation came from *Road and Track* in October who began: "It's not 'the Best Car in the World', but it's more enjoyable to drive than the vast majority of vehicles on the road today – including many sports cars". *Road and Track* felt that the Mark X had no real direct competition except perhaps for Jaguar's own 3.8-litre Mark 2. Whilst contemplating the overall size of the car (massive by UK standards), in the US the Mark X barely matched the road standard Fords and Chevrolets of the period!

Most drivers and passengers were not too taken by the front seats, in particular by the padding being in the wrong places, and it was felt that the less agile would find the high sills a major obstacle to entry and exit.

With regard to ride *Road and Track* found that no other car of the size and type gave a better combination of comfort, handling and silence.

The first English road test came from *Autocar* in November 1962, a year after the car's release to the public. The example tested had already completed over 13,000 miles but was still free of any annoying rattles or creaks from the bodywork.

Emphasising the internal size and comfort of the Mark X, *Autocar's* only complaints concerned the lack of lateral support on the front seats during spirited driving, the need for a larger rear seat armrest, and the lack of stowage facilities for odds and ends for up to five people (even the slim map pockets in the front doors on Mark IXs had been abandoned). The angle of the brake pedal in relation to the accelerator pedal was also criticised.

The sports-car stability in cornering and the superb ride convinced *Autocar's* testers of the justification for the complex independent suspension system. Some members of the team found the power steering a little low geared but generally liked it as befitting a large luxury car of this type. The report closed with the comment: "It is one of the proudest products of British industry".

Another road test would not be seen until mid-1963, when *Wheels* magazine in Australia carried out an evaluation of a borrowed demonstrator finished in Pearl Grey with automatic transmission. The tester was obviously impressed with the car but did comment that on bad road surfaces, under extreme conditions, the Jaguar could lose its balance easily. Though lyrical

Mark X and glamorous companion at showtime.

about the ego boost the Mark X gave him, he did find fault with the standard of finish, including the boot lid not fitting squarely and leaving an uneven line along the edge (a common fault on Mark Xs), the bonnet needing three or four slams to close properly (a knack even needed on later XJs), the quarterlights being stiff to operate and the front passenger seat squeaking when no one was sitting in it.

In November an extended road test was published by *The Motor*, for the first time using a manual/overdrive version of the Mark X. The opening lines reflected on the car being better looking than it really was, although credit was given to the price tag of £2,000 when the car could have cost a lot more. However, *The Motor* remained critical, remarking that the best part of the Mark X was the XK engine and that the quality of finish and fittings was "skin deep". The heating system, whilst an improvement over other Jaguars, was still archaic when compared to cars a third of the price.

The elderly manual gearbox was criticised, as it had been many times before on other models, for its vintage feel, slow synchromesh and noisiness. However, top gear performance was found to be excellent, the car pulling smoothly from 7mph to 120mph with the greatest of ease. The clutch was found to be excessively heavy, and the whole transmission required some practice to master smooth changes.

The Motor's fuel consumption figures were interesting, comparing a 25-mile motorway cruise at up to 110mph, yielding 12mpg, to a 17-mile gentle drive over country roads averaging 36mph, producing 23.3mpg. Handling was thought to be generally excellent although the poor seating again came under fire.

Visibility was criticised as the driver could not see the rear wings at all and the bulbous sides of the bodywork did not help in gauging narrow gaps (the majority of Mark Xs suffered from side scratches in everyday life). The difficulty of entry and exit was again mentioned, and the testers particularly found that when the car had been driven in the rain it was difficult to negotiate the large sills without getting one's trousers dirty.

Inside, *The Motor* found the veneering poor and the switchgear unergonomic, particularly the heater switches on the centre console (one tester actually selected the BBC Home Service instead of the heater!). Although the boot was found to be enormous, light storage inside the car was considered meagre.

It wasn't until July 1963 that the normally very critical *Motor Sport* got its hands on a Mark X for assessment, and after driving two examples of the model over some 600 miles commented: "...it enhances the already all-but-mystical ability of this Coventry manufacturer to offer a quite outstanding return, in terms of comfort, equipment and very high performance, for a comparatively modest outlay." In the report there was an interesting comparison with the Daimler Majestic Major (tested one month previously) and the tester commented on the Daimler's slight superiority in top speed and acceleration as well as its being more economical. Despite this, *Motor Sport* was not scathing in any way towards the Mark X and found it quite a superior motor car by any standards.

The final road test of the 3.8-litre Mark X came in October 1963, carried out by *Cars Illustrated*. Enthused by the exciting performance and good degree of comfort, the writer's only major criticism concerned excessive brake pedal travel after hard driving. The test car, having already covered some 11,000 miles in other hands, exhibited no signs of body rattles or age.

In August 1966 *Autocar* carried out a used car test on a 1963 Mark X for sale at £765. The car in question had had two owners and had 43,000 miles on the clock. *Autocar* suspected a higher mileage than that, but the car's only mechanical faults appeared to be piston slap when cold, high oil consumption (around 180mpp) and excessive use of coolant (over a gallon in 500 miles). Bodywork had suffered rather more severely, with areas badly resprayed and major rusting of the fuel filler hinge. Interior leatherwork was also worn, and one cannot help thinking that the car must have had quite hard use in its three years on the road. Despite all this the Mark X was considered a reasonable buy, with a depreciation of £1,257 since new – quite fair in *Autocar's* opinion at the time.

The competition

At around the time of the launch of the Mark X came many other significant new cars including the Ford Consul Cortina, the Morris and Austin 1100, the Triumph Vitesse and, perhaps a direct competitor to the Mark X, the Lagonda Rapide.

The latter, based around the Aston Martin DB4, with a six-cylinder twin cam engine and high levels of luxury and performance, offered greater prestige than the Jaguar and was handbuilt, but it was subject to many production problems, poor build quality and lack of development. Only 55 were in the end produced, the model's demise probably hastened by a very high price tag and the Jaguar's success.

The Daimler Majestic Major was still available, though now built by Jaguar, and started to appear with bits of Mark X trim before its discontinuation in 1965. It is interesting to note that even in the *Motor* and *Motor Sport* road tests of 1963 the Daimler still came out by far the fastest of all the comparison cars tested, with a top speed of 122mph compared with the Jaguar's 120mph. Acceleration was also 1½secs better to 50mph, with 20-40mph acceleration vastly superior: 3½secs compared to the Jaguar's 9secs (although the Daimler was in kickdown mode).

Rover pushed forward with a new sportier look for their upmarket 3 litre in the form of a fixed head coupé based around the standard Mark II saloon but with lowered roof line. A very attractive model with an excellent reputation, it was generally bought by the more sedate type of owner. It was competitively priced but quite slow even by the standards of the early 1960s.

Rolls-Royce and Bentley soldiered on with their Cloud/S Series, a new version of the V8 6.2-litre engine for 1963 giving superb performance. Very expensive motor cars, they nevertheless didn't handle or perform as well as the Jaguar and cost considerably more.

European competition in those days only really came from Mercedes-Benz with their 220SE model, which was more expensive, and slower, than the Jaguar, and in any event the German marque did not then enjoy the same status in England as it does today.

North American competition was as extensive as ever, with the homegrown Lincolns, Buicks and Cadillacs, but perhaps the most competitive of the era was the Ford Galaxie 500, an even larger car than the Mark X. It could out-accelerate the Mark X but lagged behind in top end performance, recording only just over 100mph. Fuel consumption was disastrous, and although very fully equipped, with electric seats and windows, cruise control and other gadgetry, the Galaxy like other American cars of the period was no match for a Jaguar in terms of handling or perceived quality.

Improving the Mark X

A continuous programme of development of the Mark X took place, some changes major and some very minor, yet all affecting and improving the car in terms of quality, reliability and performance.

All Mark Xs were available with electrically operated window mechanisms as an optional extra. The first 500 cars featured a double circuit-breaker unit, one half protecting the circuits on the left hand side of the car, the other protecting the right. Subsequent models used a refined version, a single circuit-breaker in conjunction with a second relay which also allowed the driver to retain use of his window providing the fault was not in that door. In the event of failure of the electric windows, there was a unique manual window lift system involving a winding lever kept in the boot of the car; by inserting the lever in a special aperture provided under each door, windows could be wound shut. The idea was a good failsafe but Jaguar dropped it from subsequent models.

1962

Early in the car's production, from engine nos. ZA.1001, all Mark X engines (in common with other Jaguar XK units) received a modified crankshaft rear oil seal incorporating an asbestos rope oil seal in an annular groove, the crankshaft being modified to suit. Further, from engine no. ZA.1054, in February new big end bearings with reduced running clearances were fitted.

Very early on in production Jaguar amended the design of the centre console ashtray, originally an elaborate half-moon shape which was found to be expensive and was therefore replaced by a more straightforward rectangular shape with separate pull-out knob.

In April Jaguar introduced an electrically heated rear window for all saloons to provide demisting or defrosting. The heating element, consisting of a fine wire mesh, was laminated into the rear windscreen and was connected to the wiring harness. The heating element came into operation immediately the ignition was switched on, consuming 5amps and remaining operative all the time. Wiring already existed on all Mark Xs prior to the system's adoption so a kit was made available to convert earlier models. The heated backlight remained an extra cost option.

In May, Jaguar decided to fit longer rear springs to Mark Xs. The 15.9mm increase in length allowed the removal of the previous aluminium spacer between the spring seat and split retainer. The new springs were colour coded green under part no. C.20104 and were fitted from chassis nos. 300130 rhd and 350320 lhd (although a few earlier cars may also have been so fitted). Ironically Jaguar changed the springs yet again in November, to part no. C.21280, colour coded in yellow and with the number of coils reduced to 10¼. Also in May, to avoid the possibility of condensation being drawn into the engine via the breather pipe, a 1¹⁄₁₆in

The dashboard conformed to the style started with the Mark 2 in 1959.

In terms of the use of chrome trim, the Mark X was muted in comparison with the Mark IX.

109

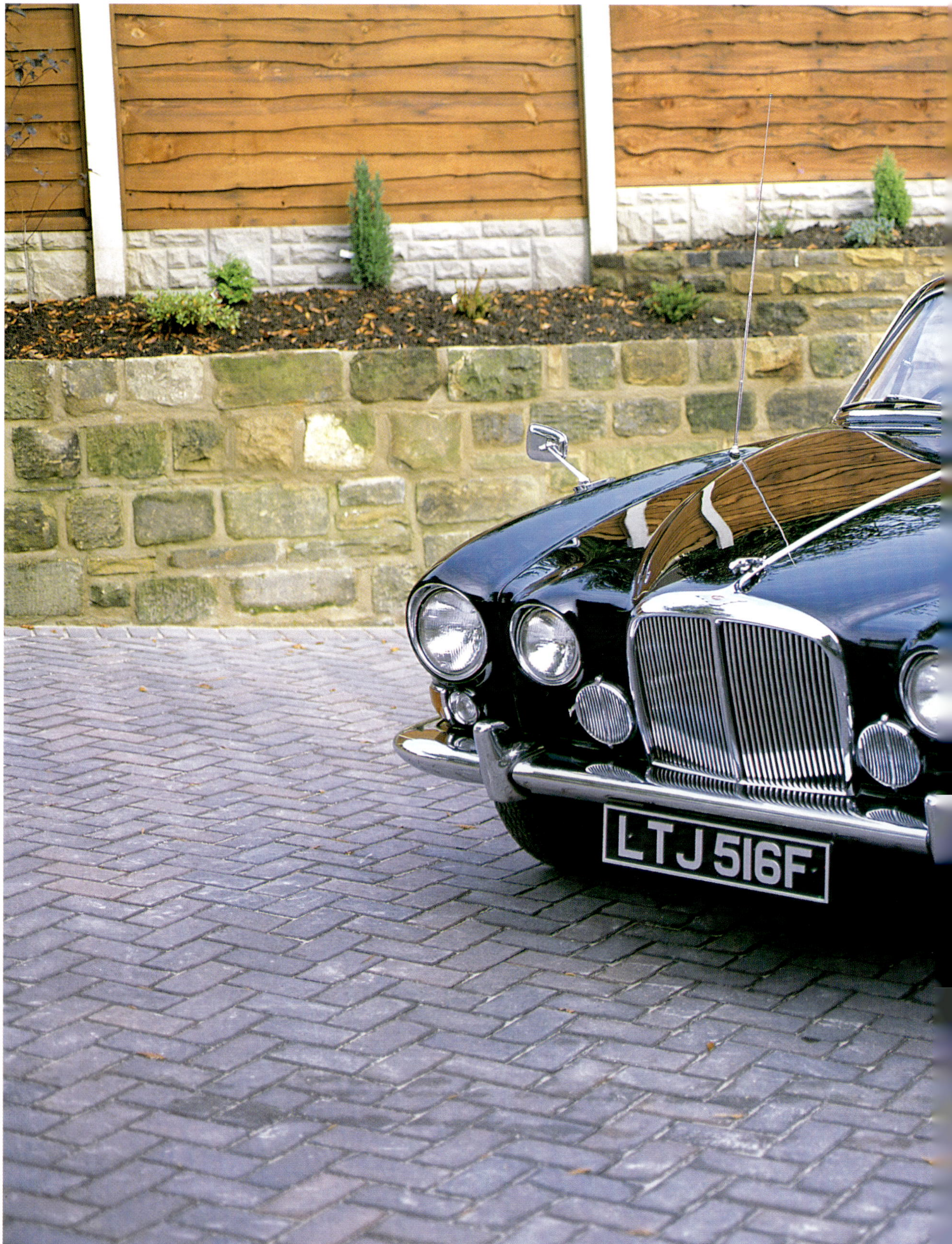
LTJ 516F

The 420G was distinguished by a central grille rib, a chrome swage line strip and side repeater indicators.

The exceptionally roomy engine bay of the Mark X. This is a 3.8-litre version. Triple SU carburettors were standard equipment on Mark X and 420G saloons.

Only the 420G badging on the bootlid differentiated the car from the Mark X at the rear.

hole was drilled in the underside of the pipe approximately 6½in inwards from the bend; standardised from engine no. ZA.1357, this modification was also carried out on earlier cars.

From engine No. ZA.1730, also in May, all engines were fitted with modified inlet camshafts with a hole drilled in the base of each cam to reduce tappet noise when starting from cold.

In July, the heater system was slightly amended, allowing the scuttle ventilator to be operated independently and consequently reducing the warm-up period of the heater in cold weather conditions. With the vacuum system, a supply tank kept a reservoir of vacuum which allowed the use of the controls up to six times after the engine was switched off. If the heater controls were left depressed after switching off the engine the scuttle ventilator would automatically close after a period and then reopen upon starting the engine with the resumption of vacuum supply.

Also in July, from chassis nos. 300471 rhd and 350973 lhd, Jaguar introduced larger brake cylinder pistons for front and rear brake calipers, the front increasing from 2⅛in to 2¼in and the rear from 1⅝in to 1¹¹⁄₁₆in. Later that year the rear caliper retaining bolts and locking tabs were replaced by spring washers and wired bolts.

Approved exterior wing mirrors were now available for the Mark X. These were made by Magnatex and supplied as standard equipment (although not actually fitted until PDI by the supplying garage). Alternatively, the mirrors could be purchased separately for fitment on the top of each front wing approximately 14½in from the headlamp lip.

In September, after complaints of exhaust rattles, Jaguar fitted all models with a bracket on the clutch housing to support the rear exhaust downpipe. At the same time Jaguar had had reports from owners of sticking throttles, causing some concern for safety. Upon inspection it was found that the toe board felts were fouling the accelerator cross shafts. Also in September, Witter made available a specially designed tow bracket for Mark Xs, which fitted through the top centre of the rear bumper. In the same month Jaguar adopted a new type of nylon petrol pipe from the pump to the tank outlet with revised banjo bolt and cork washers instead of the previous unreliable Vulkollan type.

In September, from engine no. ZA.3153, Mark X engines were fitted with new top compression rings, Maxiflex scraper rings and revised connecting rods. In October all Mark X cylinder blocks from engine no. ZA.3171 were machined to accommodate larger dowels on the left hand side. Also in October Jaguar introduced an additional opalescent paint colour, Golden Sand, replacing Bronze.

Commencing at chassis nos. 301225 rhd and 351545 lhd, in November a modified upper steering column was fitted enabling the Waso-Verkon combined ignition switch and steering lock to be fitted as an optional extra. Primarily for export markets, the switch was mounted on an extension arm attached to the steering column below the steering wheel, taking the ignition key and having four positions: Drive (Fahrt), the normal position with the ignition switched on; Garage, the off position when the key could be withdrawn with ignition off; Stop (Halt), which locked the steering column with the ignition off, also allowing the withdrawal of the key. The fourth position – Start – was never wired up as the black starter button on the dashboard remained operative.

From engine no. ZA.3972 all Mark Xs received a new engine oil dipstick with handle lengthened to clear the exhaust.

From December Jaguar fitted a new rubber coupling and driving peg (part nos. BD.20909 and BD.25088) to the electric window lift motors on the Mark X, allowing the motors to gain momentum before raising or lowering the windows. This followed complaints on the early cars that the motors appeared to be straining with insufficient power. A kit was also available to convert existing cars.

1963

Due to complaints of the front dampers rattling in their mountings Jaguar fitted a distance piece in the top damper mounting hole, and then in January starting with chassis nos. 300770 rhd and 351196 lhd new gas cell dampers were introduced.

Also in January, at chassis nos. 302914 rhd and 352060 lhd, the thickness of the rear brake discs was increased from ⅜in to ½in, which necessitated the fitting of new brake calipers and caliper mountings.

In March Jaguar asked all dealers to check early cars for slackness of the lower steering column causing a knock from the steering. Because of their findings, all cars prior to chassis nos. 305043 rhd and 352940 lhd were recalled for the replacement of spacing collar, clamp claw, sockets and clips on the lower steering column. On a similar subject drivers also complained of excessive end float in the upper steering column, necessitating the fitment of a thicker bearing with the letter "B" etched on the end face.

In the same month, at chassis nos. 302791 rhd and 352000 lhd, modified brake pedal shaft housings were fitted along with a revised servo bracket. From the same chassis number, left hand drive cars had the brake fluid reservoir bracket moved to the right front wheelarch.

In April, after numerous complaints of fuel smells around the car, Jaguar fitted a new nylon breather tube to the tanks in place of the metal breather tube. A piece of bundy tube 4½in long was inserted in the end of the nylon breather tube to protrude through an aperture in the boot floor. A revised petrol filler box drain tube was also fitted, passing through the boot floor in the front of the petrol tank. In August a pink-coloured polyurethane sponge baffle was inserted in the vent pipe of both tanks, allowing adequate breathing yet preventing petrol flowing out of the vent pipes under hard acceleration with the tanks full, and thus reducing the risk of petrol fumes.

At the same time all cars were supplied with "sealed-for-life" half-shaft universal joints.

From engine no. ZA.6622, a modified jockey pulley carrier and stop were incorporated on the automatic fan belt tensioner to limit the travel of the pulley.

At chassis nos. 305762 rhd and 352377 lhd all automatic transmission models were fitted with a larger brake pedal.

May saw the replacement of the solenoid and operating valve on overdrive models with a stronger unit, subsequently fitted to all other overdrive models under instruction from the Service Department. A new filler cap and level indicator for the brake fluid reservoir were also fitted, preventing dirt from entering the reservoir (part no. C.21889). The reservoir was amended again in March 1964 to incorporate a protective cap over the indicator plunger.

In June all Jaguar production models received a new type 2206 distributor and in August, commencing at engine no. ZA.9238, a waterproof rubber cap (part no. C.2607) was fitted to each of the plug and coil HT leads where they entered the distributor cap. Also in June, after the realisation that even minor accident damage to front wings necessitated the costly replacement of the whole front wing assembly, Jaguar introduced a nose portion, including the inner locating rims for headlamps and sidelamps, as a separate service item under part nos. 9961 (right hand) and 9962 (left hand).

In September all Mark X engines from no. ZB.1228 were fitted with improved steel specification exhaust valves (part no. C.21942). For the radiator Jaguar started to fit a 7lb pressure cap instead of 9lb, from chassis nos. 306058 rhd and 353143 lhd.

Further complaints of petrol fumes were received and this time they were tracked down to inadequate sealing between the petrol tank filler tube and the adaptor onto which the filler cap was screwed, rectified by judicious use of Araldite!

In November, after numerous owner complaints of poor water sealing, Jaguar issued a major service bulletin to enable dealers to stringently test the bodyshells and rectify sealing faults with preparations such as Flintkote 746 and Bostik 692. Also that month, from chassis nos. 304482 rhd and 352709 lhd, the electric clock fitted within the rev counter was modified, incorporating a rectifier to reduce fouling of the contact points in the clock.

By December Jaguar had adopted a "live" centre horn push button on the steering wheel for Mark 2s and Mark Xs, from chassis nos. 306489 rhd and 353182 lhd.

1964

In January Jaguar introduced Dunlop SP 205 × 14 tyres for all Mark Xs from chassis nos. 307376 rhd and 353362 lhd. This was followed by the fitment of recalibrated speedometers. The new speedometer part nos. were C.22693 (non-overdrive gearbox) and C.22691 (overdrive/automatic). By March another new tyre, the SP41 Dunlop, was being fitted, from chassis nos. 308495 rhd and 353583 lhd. It had a braced tread construction, but with a similar pattern to the previous RS5 tyre.

From engine no. ZB.3121 the sump front oil seal recess in the timing cover was modified so that the seal could be more easily replaced without removing the actual timing cover.

From chassis nos. 307612 rhd and 353412 lhd, shrouds were fitted over the halfshaft universal joints, both inner and outer.

Interestingly, the wider section SP41 tyre could not be accommodated in the boot spare wheel well without modification. A bracket welded to the side of the depression in the boot floor needed to be knocked down until it roughly conformed to the shape of the remainder of the well!

January also saw the fitment of a new radiator system to alleviate problems with air locks. Fitted from chassis nos. 307363 rhd and 353359 lhd, the new assembly had a separate header tank bolted to the top of the tube and fin radiator block and connected to the left tank by a short rubber hose. A short small-diameter hose from the top of the right tank was also connected to the header tank, acting as a vent pipe and allowing any air returned from the engine to escape into the header tank. The water pump also had to be modified.

To facilitate the fitting of the new bottom return pipe, new style horns were fitted one above the other on the left hand side of the car, with a suitably modified wiring harness.

Jaguar adopted a revised type of Borg Warner automatic transmission unit and torque converter from engine no. ZB.2682 (and a few previous cars).

These units were part numbered C.23446 and C.23447 respectively, with a prefix "P" prior to the serial numbers. The new units gave an improved gear shift.

For March 1964, commencing at chassis nos. 307781 rhd and 353456 lhd, Mark Xs were fitted with a revised type of interior headlining. Now bonded to a fibreglass board and not directly to the roof, the new headlining made for a better fit and was easier to remove. From engine no. ZB.3114 revised pistons were fitted, with a redesigned skirt panel, a chamfer below the oil control ring with seven drain holes and a relief at the bottom of the skirt.

In May all Mark Xs fitted with automatic transmission or overdrive were fitted with an 8 amp fuse in the intermediate speed hold switch (automatic transmission) and the control switch circuit (overdrive models) from chassis nos. 308244 (auto) and 308320 (overdrive) rhd, and 353519 (auto) and 353520 (overdrive) lhd.

In the same month, after further complaints from owners about "knocks" from the steering column, all cars, from the chassis nos. 308006 rhd and 353482 lhd, had Elastollan upper steering column bearings fitted. In June the outside diameter of the centre track rod tube was increased from ⅞in to 1in.

During 1964 the colour schemes available for 3.8-litre Mark Xs changed slightly as follows:

Exterior	Interior
Carmen Red	Black or Red
Cream (Old English White)	Black only
Indigo Blue	Red or Light Blue
Sherwood Green	Suede Green, Light Tan or Cinnamon
Black	Red, Grey, Light Tan or Cinnamon
Opalescent Silver Grey	Red, Dark Blue or Grey only
Opalescent Silver Blue	Grey or Dark Blue
Opalescent Dark Green	Suede Green, Beige, Light Tan or Cinnamon
British Racing Green	Suede Green, Beige, Light Tan or Cinnamon
Pearl Grey	Dark Blue or Red
Opalescent Dark Blue	Dark Blue or Red
Opalescent Gunmetal	Dark Blue, Red or Beige only
Opalescent Maroon	Maroon or Beige
Mist Grey	Red
Opalescent Bronze (reinstated)	Beige, Red or Cinnamon
Pale Primrose	Black or Beige
Opalescent Golden Sand	Black or Beige

No other developments of the 3.8-litre Mark X are recorded at Jaguar, probably because the production of a new version of the car was already under way for its launch in October 1964.

The 4.2-litre Mark X

The 3.8-litre Mark X had been a relative success and with its advanced engineering and modern styling had brought the flagship of the Jaguar marque well and truly up to date. Yet even after so short a production run – only 9129 right hand drive and 3848 left hand drive cars – the model was already becoming a little long in the tooth and could be outperformed by some of the competition.

We have already commented on the Daimler Majestic Major, for instance: highly competitive in terms of comfort, performance and prestige when compared to the Jaguar, and thus often an embarassment to the Jaguar people.

Having successfully transplanted the 2.5-litre V8 Daimler SP250 engine into the Jaguar Mark 2 bodyshell to create a saloon aiming at a slightly different market from the Jaguar versions, Jaguar decided to test the Mark X with the 4.5-litre Daimler Majestic V8 power unit, thus creating another upmarket saloon under the Daimler name, in a similar vein to the Double Six Vanden Plas version of the XJ.

The Daimler V8 engine was more compact than the XK unit and fitted easily into the Mark X engine bay. It was also lighter, which improved performance. Rumour has it that under test conditions the V8 Mark X produced a 0-100mph time of only 27secs (6secs faster than the XK-engined version) as well as a top speed of over 130mph.

Unfortunately the public were never to experience the Daimler V8 450, as further plans for the car's development were squashed by the Jaguar board, who were embarrassed that a Daimler-

4.2-litre Mark X at Earls Court Motor Show.

Mark Xs awaiting despatch to dealers, taken at Browns Lane in the 1960s.

engined version of their top-of-the-range car could outperform a Jaguar.

Interestingly, the Mark X bodyshell was also used as a development model for the forthcoming V12 engine. Again the engine, even with four overhead camshafts à la XJ13, fitted easily into the engine bay and under test showed some incredible performance figures, but alas a lot more development was needed before this twelve-cylinder engine would be ready for volume production.

However, Jaguar had been working on an improved version of the XK power unit, not only for the Mark X saloon but also for the E-Type sports car, and in October 1964 the 4.2-litre XK engine was launched simultaneously in the Mark X and the E-Type.

The cubic capacity of the engine had been increased from 3,781 to 4,235cc by increasing the cylinder bore from 87 to 92mm while retaining the original stroke of 106mm. The external dimensions of the engine block were unchanged but the bore spacings were altered. There were new pistons with a chromium-plated top ring, a tapered second ring and a multi-rail oil control ring. In effect the outer cylinders (1 and 6) were moved outwards, 3 and 4 were moved closer together, and 2 and 5 stayed where they were. Bearing positions were adjusted to suit the new crankshaft, which was stiffer and had thicker webs for increased strength, a new damper and repositioned balance weights. Water flow through the block was improved by repositioning and modifying the water channels to allow greater flow.

The increased capacity was not intended strictly to improve performance but to increase torque in the lower and middle speed range. Peak torque was raised from 261 lb/ft in the 3.8 litre to 283 lb/ft at 4,400rpm in the 4.2 engine, a 10% increase. Whilst maximum speed remained virtually the same, top gear flexibility was considerably improved.

The straight-port cylinder head from the 3.8 was retained despite the fact that the combustion chambers did not quite line up with the repositioned cylinders. However, a new one-piece cast aluminium inlet manifold with cast-iron balance pipe and integral water rail was employed, and the three SU HD8 carburettors, trunking, air cleaners, etc., were retained from the earlier engine.

All 4.2-litre Mark Xs were fitted with a Lucas 11 AC alternator operating at twice engine speed and reaching full charge at around 910rpm, with a 4TR control box for more efficient charging. At the same time a Lucas M45G pre-engaged starter motor was fitted. The ratio of starter to engine was increased from 11.6 to 12.8:1 by using a smaller pinion.

The radiator and cooling system were improved, with a revised radiator core giving a higher rate of heat exchange and providing a greater margin of safety against damage. The water pump was speeded up, with a redesigned rotor ensuring a higher flow, and this along with improved coolant circulation reduced the possibility of air locks in the system.

The cooling fan was now mounted on

a Holset modulated viscous coupling which, at engine speeds below 2,500rpm, gave a higher gearing to increase air flow. Above this speed the coupling began to slip, enabling the fan to freewheel. Jaguar claimed a saving of up to 16bhp at maximum engine speed.

A brand new power steering system called Marles Varamatic was fitted to the new 4.2 litre, developed for Jaguar by Adwest Engineering under Bendix patents. The new system worked on the principle that the ratio would vary as the steering wheel was turned, effectively reducing the number of turns lock-to-lock from 4 to under 3. In real terms this meant that the straight ahead steering position provided a ratio of 21.5:1. As lock was applied the ratio was gradually reduced to a minimum of 13.1 at full lock.

With the fitment of the alternator the power steering was now driven by its own separate vane pump via pulley and belt from the engine.

The brakes were further improved with the fitting of Dunlop Mark 3 calipers made from cast iron instead of malleable iron, the front discs having mud-shields. The pistons were of 2¼in diameter and the pads were increased in surface area, the friction material being Mintex M59, specially developed for anti-fade characteristics.

The aged and often criticised Kelsey-Hayes bellows brake servo was dropped in favour of a conventional Dunlop tandem in line suspended vacuum servo with line pressure, operated directly off the brake pedal.

The transmission options of automatic, manual or manual with overdrive were as before, but in all cases uprated units were fitted for the 4.2-litre engines. The automatic transmission now used was the successful Borg Warner Model 8, a beefed-up version of the Model 35 already used by Jaguar in other models. This transmission still gave three forward ratios but featured the D2 position allowing second gear take-offs, and had a heat exchanger built into the engine's cooling system.

Manual transmission cars were to benefit the most, with the fitment of Jaguar's brand new four-speed all-synchromesh gearbox. This was to become the standard unit for all subsequent Jaguar saloons and sports cars. The use of inertia lock baulk rings fitted to each gear provided crashproof changes and gave a light, quick change.

The gearbox casing was made of cast iron to ensure rigidity and to eliminate noise. All gears were of case-hardened high-core-strength steel, again for silence but also for maximum life. A new 10in Laycock diaphragm type clutch was employed, giving lighter load on full compression.

All these mechanical changes reasserted the Mark X's rightful status as flagship of the marque. The increased torque now gave a similar figure to the ill-fated Daimler Majestic Major with an improvement in overall performance figures:-

	4.2 litre	3.8 litre
0-50mph	7.9secs	8.4secs
0-60mph	10.4secs	10.8secs
0-100mph	29.5secs	32.9secs
Top speed	122mph	120mph
Standing ¼ mile	17.4secs	18.4secs

According to contemporary road tests even the fuel consumption had improved, from an average of 13.6mpg to 16mpg.

Externally the Mark X was totally unaltered except for the discreet "4.2" badging on the boot and the fitment of very slightly altered road wheels to eliminate fouling the brake pipes.

Jaguar introduced additions to the exterior paint finishes on Mark Xs with Warwick Grey (Pinchin Johnson manufacture) and Dark Blue (a non-metal-

Revised interior showing the new heater controls on the centre panel.

lic colour) replacing Indigo and Pale Primrose (both the latter from ICI). It would appear that hardly any right hand drive Mark Xs were produced in Primrose, the colour being reserved for export models, mainly with a black leather interior.

Internally the 4.2-litre Mark X was indistinguishable from its predecessor except for a new heating and ventilation system planned for ease of operation and flexibility. The first advantage of the new system was to allow both front passenger and driver individually to select their own temperature control and direction of air flow. This was controlled by four slider controls now situated within a veneered panel set above the centre console within the parcel shelf area. Air direction was controlled by the lower of the two controls, temperature by the upper. The fan was still operated by the toggle switch on the centre dash panel but the three air/heat control push buttons had now moved to the centre of the parcel shelf edge instead of being mounted on the centre console itself. In April 1965 Jaguar provided a conversion kit costing the princely sum of £7. 10s. to convert the vacuum heater system to a semi-manual operation by the use of control levers and linkages. This involved cutting the trim and making modifications to the console, but allowed intermediate heat settings and the air temperature to passenger and driver outlets to be controlled independently.

Another minor change at this time involved the screenwasher system, which was changed in keeping with all other Jaguars of the period. The old style glass water reservoir and Lucas 2SJ vacuum unit were discarded in favour of a one-litre high-density polythene container and the Lucas 5SJ electrically operated unit, comprising a small electric motor driving a centrifugal pump through a three-piece Oldham type coupling. Operation was via the same dash-mounted switch although the system only remained operative so long as the switch was pressed, the motor cutting out if the water reservoir was empty. This modification came into effect from chassis nos. ID50548 rhd and ID75374 lhd. The access panel in the right hand side of the gearbox tunnel to aid removal of the starter motor was deleted at the same time.

The new 4.2-litre Mark X was released to the public at the Earls Court Motor Show in October 1964 at a price of £1,783 plus purchase tax of £373. 0s. 5d., making a total of £2,156. 0s. 5d. in non-overdrive form. The manual/overdrive transmission version came out at £1,833 plus £383. 8s. 9d. purchase tax, making £2,216. 8s. 9d. The Borg Warner automatic transmission car cost £1,895 plus £396. 7s. 1d. purchase tax making £2,291. 7s. 1d.

At the close of 3.8-litre Mark X production the price for the manual/overdrive model stood at £2,082. 10s. 5d., making the 4.2-litre version £134 dearer (with tax). The improved specification far outweighed this modest increase in price.

What the Press said

Motor carried out the first official road test of the 4.2-litre Mark X in October 1964 on an automatic transmission version, stating that it was more rewarding to drive and well worth the extra money. The steering and brakes were particularly praised.

In August 1965, John Bolster tested a manual transmission Mark X for *Autosport* and was obviously impressed. Considering himself lucky to have the opportunity to drive a manual transmission model, he achieved a timed maximum speed of 128.5mph.

Bolster found the power assisted steering excellent and uncannily capable of controlling the car very precisely on wet roads. Braking was also found to be well up to the performance of the car.

Wheels magazine carried out brief road impressions in 1965, considering the Mark X still unbeatable value for money. The Varamatic power steering was though one of the best systems in the world, responding to the slightest movement but with adequate feel. The brakes never faded despite severe use and even the handling was improved with the new Dunlop tyres.

Later that year *Autocar* carried out a road test, finding the manual version slightly slower – 0-60mph 10.4 secs (9.9 secs auto), and 0-100mph 29.5 secs (27.4 secs for auto) – but with a marginal improvement in top speed with the aid of overdrive to 122.5mph. The all-synchromesh gearbox was considered a major improvement although movement between gears was still thought to be too long.

Autocar reiterated its earlier comments on the Jaguar's superb value for money and felt that the 4.2-litre Mark X was perhaps the most remarkable Jaguar of all.

The competition

As one of the motoring magazines of the period put it, the Mark X 4.2 litre was alone in the luxury car sector at its

Mark X production line at the Browns Lane factory.

price. Although the UK market for new cars had increased dramatically to a record one million, new or existing models in the luxury saloon range were thin on the ground. Luxury sporting cars like the Aston Martin DB5, Bristol 408 and even Gordon-Keeble were available, but these were hardly of the same type as the Mark X. Rolls-Royce and Bentley were still producing their V8-engined saloons in four-headlight form, but at a substantially higher price than the Jaguar.

In the mid-range sector, Rover were still producing their 3 litre alongside a brand new 2000 model to match the Triumph 2000, and BMC were producing their prestige Princess R, based around a Rolls-Royce six-cylinder engine. The latter lacked the speed and acceleration of the Jaguar and used a standardised bodyshell borrowed from Wolseley and Austin models.

Only Jaguar themselves had anything to offer in the way of home-grown luxury competition. Alongside the Daimler Majestic Major (still a worthy competitor to the Mark X), the company introduced the new S-type, which for the first time offered a mid-range choice between the Mark X at one end of the scale and the Mark 2 at the other. Providing excellent performance and quality, the S-type gave the Jaguar devotee a better ride than the Mark 2 as well as benefits in price, size and handling over the large Mark X.

At the Earls Court Motor Show, alongside the new E-Type, S-type, the established Mark 2, the Daimler V8 and the Daimler Majestic Major were two Mark Xs, a manual/overdrive car in opalescent Silver Grey with grey interior and a Golden Sand automatic model with tan interior, both featuring Dunlop SP41 whitewall tyres.

Continued development

1965

In June 1965, 4.2 litre Mark Xs from chassis nos. ID.50608 rhd and ID.75415 lhd were fitted with modified track rod ends, pre-packed with lubricant and therefore requiring no maintenance.

In September, due to changes in legislation in North America, Jaguar had to fit all Mark Xs with hazard warning lights, which they did by adapting the four direction indicators to flash simultaneously via a toggle switch mounted on a sub-panel under the dashboard. From engine no. 7D.52037 a hydrostatic clutch-operating slave cylinder was fitted, allowing normal clutch wear to be automatically compensated for.

In November, Dunlop introduced a special radial ply tyre with a winter tread pattern, known as the Weathermaster SP44, for use on the rear wheels of Mark Xs in adverse weather conditions.

1966

In April all Mark Xs were fitted with a revised direction indicator switch from chassis nos. ID.51955 (rhd) and ID.76088 (lhd). The new switch trip mechanism operated through a freely

Final polishing at the factory.

rotating castellated nylon striker ring in direct contact with the striker and was attached to the lower column, giving a more positive action to the switch.

Also in April the heated backlight on Mark Xs was amended by fitting a separate on/off switch, a warning light and a relay with resistance included in the electrical circuit. This was effective from chassis nos. ID.52838 rhd and ID.76425 lhd. The switch and light were situated on the dash panel adjacent to the brake fluid warning light. The heating element only became activated when the switch was "pulled" and the light lit up (the light automatically dimming if the sidelights were switched on).

At the same time a Delaney Gallay air conditioning system became available as an extra-cost option for 4.2-litre Mark Xs. This necessitated a modified radiator, coupled with a larger diameter fan assembly, which were subsequently standardised for all Mark Xs up to the end of production. This modification became effective from chassis nos. ID.51950 rhd and ID.76088 lhd. The air conditioning unit was mounted in the boot and drew warm air from the car interior through ducts in the rear parcel shelf, passed the air through an evaporator to be cooled and then back to the passenger compartment. The compressor for the system was engine mounted and driven by belt and pulley. The system was controlled from a black crackle-finished panel under the centre of the dashboard above the centre console. It had a simple on/off switch to activate the clutch on the engine-mounted compressor, a two-position switch to operate the fans and a thermostatic temperature controller. It was priced at £275. 10s. and not many Mark Xs were fitted with it.

At this time the range of standard exterior paint finishes for Mark Xs amounted to:

Exterior	Interior
Opalescent Silver Blue	Grey or Dark Blue
Opalescent Silver Grey	Red, Dark Blue or Grey
Pale Primrose	Black or Beige
Sherwood Green	Suede Green, Light Tan or Cinnamon
Carmen Red	Black
Opalescent Maroon	Maroon or Beige
Old English White	Black or Maroon
Warwick Grey	Red, Light Tan or Dark Blue
Black	Red, Grey, Cinnamon, or Light Tan
Dark Blue (non metallic)	Red or Grey
Opalescent Dark Green	Suede Green, Beige, Cinnamon or Light Tan
Opalescent Golden Sand	Red or Light Tan

By December 1966, after continuous complaints of brake squeals, Jaguar recommended the fitment of new Mintex M.74 friction pads replacing the previous Mintex 59 material used.

Total 4.2-litre Mark X production only amounted to 3729 right hand drive and 1960 left hand drive cars between October 1964 and December 1966.

By 1966 the price of the manual/overdrive 4.2 litre Mark X was £1,870 plus £391. 2s. 11d. purchase tax, making a grand total of £2,261. 2s. 11d., without extras such as the electrically heated rear window, radio, etc. The Mark X remained competitively priced right to the end and although Jaguar would have liked to have had an entirely new model ready to replace it this was not to be.

Instead the company opted for a revised package to be known as 420G, the next generation of Mark X, discussed in the next chapter.

Last of the line:
The 420G

Styling drawing for the 420G showing a more pronounced centre grille and smaller outer headlights.

By autumn 1966, for the 1967 model year, Jaguar was running a revised model numbering system; the Mark 2s were amended in specification and re-designated 240 and 340, a brand new intermediate model between the S-Type and Mark X called the 420 came out, and the Mark X itself received a facelift, with re-badging to model designation 420G (presumably 420 denoting the engine size – 4.2 litres, and the "G" indicating Grand – A Grand Touring Version of the 420!).

In most published references to the Mark X range, the 420G is unfortunately dismissed as a hastily prepared final fling to improve sagging sales figures until the introduction of the XJ6, and as being aimed at the North American market, with its brash two-tone paint finishes and extra brightwork. Despite such comments the 420G was still Jaguar's flagship, and it sold better than the 4.2-litre Mark X which it replaced. Jaguar's own brochure enthused: "...for the utmost in high performance prestige motoring the Jaguar 420G is a most spacious five seater saloon with a comprehensive specification which offers luxurious comfort, lavish appointments, superbly smooth power and consummate safety assured by splendid acceleration, perfect road-holding and positive stopping power... in short, Jaguar engineering ensures a degree of perfection unsurpassed in modern motoring".

Introduced in October 1966, to run concurrently with the 4.2-litre Mark X until stocks of that model were

The 420G's swage strip and new hub caps made an appreciable difference to the Mark X's looks.

When supplied in a two-tone paint scheme, the 420G's chrome swage strip was omitted.

exhausted two and a half months later, the 420G retained exactly the same mechanicals and bodyshell, so, in effect, all the changes were cosmetic. At the front, the radiator grille gained a thick chrome vertical centre rib. Indicator repeaters were added to the front wings and plated swage line strips ran the whole length of the body. The hub caps were now of cleaner appearance, with black plastic inserts in the centre containing the Jaguar head motif. Simple push-fit chromium-plated Rimbellishers complimented the 14in wheels painted body colour. At the rear the only tell-tale was the "420 G" badging on the boot lid.

Perhaps the most noticeable change was the availability of two-tone paintwork, which when specified had the distinct effect of making the car look even longer although it did little to disguise its bulging flanks. The range of standard colour schemes (both solid and two-tone) upon introduction was:

Exterior	Interior
Cream	Red, Light Blue, or Dark Blue
Beige	Red, Suede Green, Tan or Light Tan
Warwick Grey	Red, Dark Blue or Tan
Willow Green	Suede Green, Tan, Beige or Grey
British Racing Green	Suede Green, Beige, Light Tan or Tan
Dark Blue	Red, Light Blue or Grey
Black	Red, Grey, Tan or Light Tan
Golden Sand	Red or Light Tan
Opalescent Silver Blue	Dark Blue or Grey
Opalescent Silver Grey	Red, Light Blue, Dark Blue or Grey
Opalescent Maroon	Maroon or Beige

Two-tone colour schemes

Black over Opalescent Silver Grey	Red, Grey, Tan or Light Tan
Black over Golden Sand	Red, Grey, Tan or Light Tan
British Racing Green over Willow Green	Suede Green, Beige, Tan or Light Tan
Dark Blue over Opalescent Silver Blue	Red, Light Blue or Grey

Inside, there were new aerated leather centre panels in the seats, and the shape and style of the seats were slightly altered to improve lateral support. Woodwork, trim and switchgear were all exactly as on the previous model except for the following. The dashboard cant-rail top was now trimmed in two separate panels of black vinyl, padded as a token gesture towards crash protection. In the centre of the reshaped cant-rail was a square electric clock, replacing the clock in the rev counter face. The rev counter obviously also changed accordingly and was now ignition impulse driven as on other Jaguars of the period.

The automatic transmission lever on the steering column was now a longer, cranked stalk with a black knob, making it look more like a control lever and less like an additional indicator stalk!

Under the bonnet, the 420G had black-painted, ribbed camshaft covers, standardised on all Jaguars from this time on. The new model was mechanically identical to the Mark X in other aspects except for the upgrade in automatic transmission to Borg Warner Model 8, which gave D1 and D2 drive positions.

Launch prices for the 420G were quite attractive, the standard manual (without overdrive) version priced at £2,237 including purchase tax. This was actually slightly cheaper than the Mark X 4.2-litre and actually only £39 dearer than an automatic transmission version of Jaguar's new medium sized saloon, the Daimler Sovereign. With overdrive the 420G cost £2,300 including purchase tax and in automatic form £2,380.

What the Press said

As the introduction of the 420G didn't actually create a stir in the motoring world it was some time before magazines published assessments and road tests. *Motoring World* was the first to test the car, in April 1968. Although the changes were felt advantageous to the car, the front wing indicator repeaters were not thought necessary as the indicators had originally been fitted as wrap-around units, making them instantly viewable from the side. There was also, surprisingly, criticism of the rear compartment legroom!

Motoring World did, however, compliment the Jaguar's power steering and brakes, as well as overall performance, even if in the lower speed ranges the XK engine seemed noisy. In all, the 420G was considered a big improvement over the Mark X, but some of the erroneous comments made suggest the writer was comparing the 420G with the earlier 3.8-litre Mark X.

The Australian *Modern Motor* magazine looked at the 420G in September 1968 and the opening statement said it all: "Grace, pace, elegance – and only $8800! The Jaguar 420G offers just about everything any owner would want". On a 1,420-mile road test (alongside a Toyota 2000GT) the Jaguar proved highly impressive. In a 30-knot crosswind the Jaguar easily outperformed the "high performance" Toyota sports car in acceleration.

Compared with the previous road-test, *Modern Motor* staff found the legroom and quality of finish quite excellent. They did, however, severely criticise the safety aspects of the steering lock (fitted to export models). This allowed the steering to remain locked when the car was in motion, as the Jaguar was still fitted with the conventional key ignition and starter on the dashboard. The power steering and handling were found to be exemplary despite the car's bulk, and there were enthusiastic comments on the car's ability to handle S-bends at 90mph without trouble. Criticisms, apart from the steering lock, were minimal. The wipers were inadequate at speed, the

Indicator repeaters at the head of the chrome trim were featured on the 420G as a token towards better road safety, although the car already had wrap-around indicator lenses.

420G dashboard showing revised clock position and padded vinyl rails.

Many Mark X and 420G models were fitted with Jaguar's own seat belts made up by Britax. Perforated leather was used on 420G seats.

Revised and simplified hub caps for the 420G, subsequently used on 240, 340, 420, Daimler V8 and early XJ models.

heating system lacked capacity and the interior mirror vibrated badly.

1988 saw a secondhand 420G being assessed by Robert Davies in *Classic Cars* magazine. The car concerned had belonged to Jaguar's own engineer, Claude Baily. It was one of the last 420Gs, had been a Press car initially, and was used extensively by the factory until taken over by Mr Baily upon his retirement in 1972, with around 36,000 miles on the clock. Finally sold at just under 50,000 miles, the car as tested by *Classic Cars* was still in fine original condition. Despite its being an old design, the tester was impressed by the car's smoothness, ride and handling capabilities.

In March 1990 *Your Classic* magazine also looked closely at the 420G along with the earlier Mark X models. The tester was obviously very impressed with the Jaguar despite its bulk.

In 1991 *Classic and Sportscar* compared a good example of the 420G with the contemporary Bentley S3 saloon. Working on the premise that the Jaguar was a bargain when new compared with the Bentley, the magazine wanted to evaluate the two and determine which was the better car.

Findings were that the Jaguar handled superbly and was, in the main, a better driver's car than the Bentley, but the latter finally won on sheer build quality and finish. However, the 420G was considered by *Classic and Sportscar*'s writer to be a really good car, better than it had ever been given credit for.

The competition

By the time of the 420G's introduction large cars were a little out of fashion, as medium-sized cars like the Triumph and Rover 2000, and even Jaguar's S-Type and 420, were now the executive's choice. British-made competition to the gigantic 420G had virtually disappeared and at the time of the car's launch, a famous name, Alvis, had already dropped from the scene.

Jaguar's own in-house competition, the Daimler Majestic Major, was also just going out of production, which meant that apart from the small-scale specialists like Bristol, Jensen and Aston Martin, the only other competitors came from Rolls-Royce and Bentley in the form of their brand new Silver Shadow and T Series. Highly praised upon their release in 1966, the Crewe-built cars offered a similar turn of speed, fuel economy, ride and comfort to the Jaguar in a similarly proportioned bodyshell. However they were significantly more expensive and didn't handle nearly as well as the Jaguar.

420G fitted with a full-length Webasto folding roof. This shot conveys a good impression of the bulging flanks of the car.

A rare colour publicity photograph of the 4.2-litre Mark X, which was indistinguishable from the 3.8-litre model.

A range of two-tone colour schemes was available on the 420G.

Front compartment of a manual transmission 420G, showing the padded dash rails and perforated leather seat panels.

The rear compartment was in the sumptuous Jaguar tradition.

Production changes

1967
Little in the way of modifications took place over the final years of the 420G production but, in March 1967, oil consumption was reduced by the fitting of inlet valve guide oil seals from engine no. 7D.55850. In July, from chassis nos. GID.53869 rhd and GID.77097 lhd all cars were fitted with front and rear seat belt anchorages as standard.

1968
The range of colour schemes for 420G models was amended to the following:

Exterior	Interior
Cream	Red, Light Blue, Dark Blue
Warwick Grey	Red, Dark Blue, Cinnamon
Ascot Fawn	Red, Beige, Cinnamon
Sable	Beige, Grey, Cinnamon
Regency Red	Beige, Grey
Willow Green	Suede Green, Beige, Grey, Cinnamon
British Racing Green	Suede Green, Beige, Cinnamon
Light Blue (non metallic)	Dark Blue, Grey, Light Blue
Dark Blue (non metallic)	Red, Light Blue, Grey
Black	Red, Grey, Cinnamon

Two-tone colour schemes

Black over Ascot Fawn	Red, Grey, Cinnamon
Black over Warwick Grey	Red, Grey, Cinnamon
Dark Blue over Light Blue	Red, Light Blue, Grey
British Racing Green over Willow Green	Suede Green, Beige, Cinnamon

At the beginning of 1968 a replaceable fuel filter had been added to the system and in July, for 9:1 compression ratio engines only, Hepworth & Grandage pistons were fitted.

A few cars later, Jaguar moved the engine mountings, transferring their fixing from the bodyshell to the front suspension crossmember beam, which eliminated vibration from the engine being carried through the shell. No alterations were necessary to the crossmember but a revised cylinder block was used which, because of these mountings, was not interchangeable with earlier engines.

From chassis nos. GID.54196 rhd and 77279 lhd, a modified Saginaw power steering pump was fitted, with different shaft and pulley mounting, also affecting the jockey pulley mounting bracket.

From July all 420G dash-mounted electric clocks were fitted with domed nuts instead of conventional brass knurled nuts. The new nuts had to be removed by a special tool, eliminating the problem of pilferage that had been experienced.

In the same month, from lhd German market chassis no. GID.77569, a side light bulb was included in the outer headlamp unit. The side lamp units were still retained, but not wired up. Also, for all cars from chassis nos. GID.55267 rhd and GID.77609 lhd, all Lucas ignition coils had push-on HT terminals.

In August, Signal Red (a bright red) became available, although it was little used on the 420G model. Also, a two-tone scheme of Sable Brown over Cream was available for a limited period.

The demise of the 420G had been on the cards for some considerable time, and with the introduction of Jaguar's car for the 1970s, the XJ6, the old 420G was surplus to requirements. However, although the company replaced all other models in favour of the XJ6, the 420G actually soldiered on until August 1970, outliving the other models. Several British "J" plated cars were registered, although some of these were heavily discounted by dealers to clear.

1969

In March, due to a standardisation of parts with the new XJ6 model, all remaining 420Gs built featured a Girling type 100 Supervac brake servo unit, effective at chassis nos. GID.56264 rhd and 77837 lhd. Again due to a shortage of earlier parts, many earlier models were subsequently fitted with the Supervac when replacements were needed.

In April, 420Gs were fitted with a new three-position ignition switch with a position to the left of "off" for the operation of auxiliary accessories, like radio, with the engine switched off. This modification became effective from chassis nos. GID.56575 rhd and GID.77881 lhd.

June saw the replacement of the cylinder block drain tap with a drain plug at engine no. 7D.59989, and in August the engine number was moved to the crankcase bell housing flange on the left hand side of the engine for ease of checking, effective at engine number 7D.60230.

In October, the mercury cell-powered electric dashboard clock was replaced by one powered by the electrical system of the car itself. The new clock was part no. C.32437 (old mercury cell-powered, C.26989) and this became effective at chassis nos. GID.57384 rhd and GID.77983 lhd. Due to lack of availability of the earlier battery-powered clocks, dealers were requested to amend the wiring of existing cars to suit the new clock.

In November, in common with other XK-engined Jaguars, the 420G received new camshafts with redesigned cam profiles to give quieter valve operation over a wider range of valve clearances, and longer periods between tappet adjustments. This modification affected 420G engines from no. 7D.60699.

1970

By March, nearing the end of 420G production, further standardisation took place, so the very last models had the same exhaust manifolds as emission cars, from engine no. 7D.60778.

The ten-year production period of the Mark X had apparently come to an end; a controversial period in Sir William Lyons' and Jaguar's story during which it could be argued the model never received the praise it deserved. A major departure from anything William Lyons had ever created before, the Mark X still proved a commendable adversary to the competition over the period.

The total production for the bodyshell was made up of:

	UK	Export
3.8-litre Mark X	7607	5775
4.2-litre Mark X	2828	2291
420G	3435	2304
Limousines	42	

The grand total of 24,282 was just over half the number of Mark VII–IX cars produced in a similar period, but one could say that the Mark X had to share its market with a larger number of other Jaguar models than the previous models had.

Mark X/420G production was small by any standards, even when compared with Rolls-Royce, who managed to turn out nearly the same quantity (21,205) of Silver Shadow/T Series cars from 1966 to 1976. So the success of the model cannot be judged on the number made but on the merits of the car itself. Here the Mark X/420G must surely be classed as one of Jaguar's successes. Its well-engineered design was to pave the way for the XJ, and it must not be forgotten that a vast number of the 24,000 cars produced still exist today due to the sheer strength and sound engineering of the design.

The Mark X and 420G today

John Rundle (owner of a Mark VII as well) has owned his 1964 Sherwood Green Mark X for over four years, originally buying it from a dealer who used to maintain it on behalf of its first lady owner in the Southend area. With only 54,000 miles on the clock the car is still totally original except for minor paint work, apparently to rectify dullness of paint over the years. The suede green interior trim is also immaculate, as is the rest of the car even down to the original Hardura spare wheel cover, which has not even been discoloured by the tyre.

When originally bought the Mark X had an MoT certificate but John needed to carry out major work on the steering and brakes to rectify problems arising from lack of use over the years. He now finds the Mark X extremely practical, and has used it for family transport on many holidays, including two trips to Le Mans in France to watch the 24 hour race. His family also likes the Mark X because it is commodious and modern in feel compared with his Mark VII and other classics that he has owned in the past. The only problem in regular use is the sheer size of the car. Being bulbous sided it is difficult to park and John is always aware that it could easily get damaged in the street by careless individuals who don't realise how wide the car is.

Stewart Martin also owns an early 3.8-litre Mark X, his first classic and first Jaguar. He really wanted an S-type or 420, and was not at all interested in this larger Jaguar saloon until he actually saw one and drove it. He quickly realised the hidden qualities of the car, particularly as the example concerned seemed very original and honest and was available for significantly less than he was expecting to pay for an equivalent S-type.

The car is finished in Opalescent Silver Grey with red interior, and seems to have led a sheltered life in the hands of an older gentleman. When purchased the car needed an MoT certificate and remedial work to one sill, the jacking points and the brakes, after which Stewart set to work T-cutting the paint to bring it back to pristine condition.

The car has stood the test of time well and drives like new. Stewart feels confident that the car could take him anywhere without fear of breakdown, and he finds it substantially built compared to many later cars including Jaguars.

For the later 420G version of the Mark X I approached Ken Shipley, who has been enthusiastic about Jaguars and classic cars in general for many years. Having owned numerous Jaguars he was interested in something out of the ordinary and unusual — enter the 420G. In the motor trade himself, Ken heard of an "old Jag" which for many years had been sleeping in a garage near Manchester. Having made an appointment to see the dusty car, Ken was surprised to find that it was a 420G, not the wire wheeled 3.8-litre Mark 2 he was hoping for. The 420G was not a model that Ken felt had investment potential, and indeed he considered it ponderous and ugly to the extreme! After due deliberation and various offers he decided not to buy the car as it was too expensive despite the genuine 35,000 miles on the clock and apparently excellent condition. However, after several pestering telephone calls from the seller the car, a rare manual/overdrive version, finally became Ken's.

After fitting a new battery and freeing the sticking clutch, he was able to drive the car home on trade plates and set about the 420G to see exactly how good it really was. The original Embassy Black paintwork was excellent except for stone chips and minor scratching on the sides. So good was the overall condition of the bodywork and chrome that Ken decided to invest in a bottom half respray, which has now put the car in as good a condition as new.

Internally, the car is also in mint condition, even lacking the usual driver's seat sag and scuffs. The carpets are totally intact and the rear seat still hides under the polythene coverings fitted at the Jaguar factory. The headlining needs a good clean but other than this the car is in magnificent condition.

Mechanically the car sounds and drives like a genuine 35,000 mile example, even sporting the original Dunlop tyres. Despite its bulk, Ken and his wife really enjoy taking it to shows, where it not only obtains admiring looks but regularly takes awards for its virtually concours condition.

I have also owned two 420G models (both finished in black). My first example (actually registered with the 420G number) was a second Jaguar to my proud Mark 2 saloon. An ex-wedding hire car, it suffered from a well-worn interior (with under-felt full of confetti), some rust in the rear wheelarch areas, sills and jacking points, and an oily engine.

Nevertheless, after major expenditure on an engine rebuild and the eventual fitment of new rear wings and sills the car was very enjoyable. It was certainly a long-legged tourer. I remember several trips to far-off Jaguar events in Scotland (via the A74) and the south (via the A1) when the 420G regularly managed 17mpg at a steady 70mph, a very relaxing motor car to drive indeed.

The car was quiet and comfortable and had an air of opulence and superiority that the smaller 1960s Jaguars (including the Series 1 XJs) could not get close to. It was only when one parked near an XJ that one realised how bulbous the 420G design really was, and it was only when one drove an XJ6 that it became clear that the latter was quieter and handled better.

And finally... the Lotus Seven driver's view of the mighty 420G, widest production car ever made in Britain (except for Jaguar's own XJ220).

Mark X
Specials and One-offs

In this chapter we look at some of the better-known as well as some of the more unusual versions of the Mark X and 420G models. We have already mentioned the development car, or should I say cars, as a total of three (chassis nos. ID.50002/3/4) were used to test the V12 engine. Since that time at least one person in England has successfully grafted in an XJ12 5.3-litre engine. After all, there was always room for the unit and the bodyshell could cope easily with the added power. It is perhaps surprising that more people have not modified their cars in this way and that specialists have never offered this rather desirable conversion.

Apart from the above, and the fact that one Mark X was fitted with a Daimler radiator grille on a trial basis, the only other development that took place was towards the end of the car's production life. The rarest of the production versions was the Limousine. Originally developed from the 4.2-litre Mark X in 1965, the model was made to order to meet requirements for a chauffeur-driven luxury vehicle without the need to resort to the carriage-trade type of limousine. Chassis number designations all included "Mark XB", a throwback to the earlier Mark

A V12 engine (and a plethora of air horns) in a Mark X engine bay.

The division of the Mark X limousine seen from both sides. Note the bench front seat.

Sir William Lyons' Mark X limousine had a cocktail cabinet.

VIIIB designation given to limousine versions of the Mark VIII/IX. Utilising the conventional bodyshell, externally the Mark X limousine was indistinguishable from the saloon, not even sporting extra adornment or special badging. Mechanically the car was also identical.

Inside, there was a factory-installed soundproofed division with sliding glass partition and provision for a clock and radio for the benefit of rear seat passengers, along with the usual picnic trays (here identified as writing tables) recessed into the bottom section.

The fitment of the division necessitated a revised front seat arrangement which did restrict its movement somewhat. Other minor changes included the fitment of reading lights for the rear compartment and simple chromium-plated door pulls on the front doors replacing the usual trimmed armrests of the saloon.

Heating and ventilation systems remained unchanged and an air conditioning system was available. It was also possible to have other optional fitments, including cloth upholstery in the rear compartment, a cocktail cabinet, an intercom, and a rear window blind.

Limousine production continued into the 420G era, the only amendments being the same external and internal styling changes as for the 420G. Initially introduced at a price of £2,393 in standard transmission form, or at £2,455 with overdrive, and £2,533 with automatic transmission, a total of 15 right hand drive and three left hand drive 4.2-litre Mark Xs, 14 right and 10 left hand drive 420Gs were made. Colour schemes were sombre and limited to black, dark blue, grey and white. Optional extras included a rear compartment radio with extension speaker at £50, an air conditioning unit at £276, electrically operated windows at £58, and wing mirrors at £3.

William Lyons actually had one of the limousines, chassis no. ID.51762, a a black 4.2-litre Mark X registered FRW 169C, which was separately fitted out with a cocktail cabinet and many other

The stretched Mark X built by David Hannah – really a lot of car.

sundry items including personal dictation equipment. Upholstered in black leather in the front and grey West of England cloth in the rear compartment, the car was eventually replaced in 1968 by a 420G version, chassis no. GID 55943, registered MKV 77G and finished in dark blue with leather upholstery throughout. It is believed that other limousines went to various embassies, to businessmen and for civil duties, one export model even being shipped out to Hong Kong for the army. A few limousines fortunately still survive.

The first experimental Mark X was numbered 300001 and the original demonstration model, numbered 300005, was registered 6100 RW. This car was followed by the two 1961 Motor Show cars, numbered 300006 and 300007. One of the endurance test cars which knocked up something like half a million miles was chassis number 300050 registered 5437 RW, which was scrapped afterwards. Another development car, numbered 300051 and registered 5438 RW, was also later scrapped. The other known experimental car, which was used for testing various types of transmission, was number 351854.

The Mark X and 420G models have to some extent been the underdogs of the Jaguar saloon car range. Not many therefore have ever been converted or modified or even used as direct donor packages for replicas and kit-cars, although the engine (particularly with the three-carburettor head) and the rear suspension were, and still are, popular for such projects.

One of the most unusual conversions was carried out on a 1964 3.8-litre Mark X with automatic transmission by David Hannah from Middlesex. The body was significantly "stretched" to make the car a six-door limousine. Literally cutting the Mark X in half (which took a day) Dave welded in a new centre section cut out from a scrap Mark X. Two miles of welding wire were used and for reinforcement 12 gauge metal centre sills were welded into place, plus four original Jaguar Mark X sills, and even stronger oversills welded on top of those. The finished sills were covered in black vinyl to match the roof. To finish off the welding a massive chassis was fitted underneath the car to ensure torsional strength. Completely new centre doors, hinges, locks, etc., had to be made up and the car took a total of ten litres of Diamond White paint to finish. All sorts of special parts had to be made up including a split propshaft, exhausts, and brake pipes. David replaced the transmission with an XJ12 unit along with XJ type 15in

Graig Hinton produced several Mark X/420G based convertibles like this one.

The only known Mark X pick-up, built by a sculptor to transport his work.

chromium-plated wheels, hub caps and tyres. The interior was completely refurbished in red leather with velour headlining, and twin occasional seats taken from a Daimler Majestic Major were installed.

In the late 1970s Graig Hinton from Hinckley in the Midlands (well known for his Jaguar racing exploits in both the UK and the USA, but now sadly passed on) came up with the idea of turning the Mark X/420G bodyshell into a true convertible. He tried many experiments, some of which looked hideous. But by the early 1980s development had enabled Graig to come up with a viable conversion. The conversion meant cutting off the roof and window frames, welding up the rear doors and adding extra structural strength to the sill and underfloor areas. New pillarless windows were fitted (either manually or electrically operated) although the rear wind-down quarter windows had to be manually operated due to a lack of space to fit electric mechanisms. The specially constructed hood and frame, however, were electrically driven, with giant rams from the Rolls-Royce Corniche. Unfortunately, quite a few of the dozen or so cars produced did not feature a proper hood as this development came late in the production of the cars. The cars produced were based on Mark X and 420G shells and were finished in various colours such as white, pastel blue, two-tone blue, two-tone green, most being exported to Europe and the USA.

To my knowledge, only one person (a sculptor) has ever converted a Mark X bodyshell to a pick-up vehicle. The finished result is a practical delivery vehicle because of the enormous size of the car.

Daimler DS420
the Mark X in New Clothes

After the Daimler takeover by Jaguar in 1960, the Majestic Major saloon managed to reach production in November of that year and stayed in small scale manufacture until the mid-1960s, with the final total number of cars 1,154. Shortly after production commenced, the long-wheelbase version, designated DR 450, came onto the scene at a very attractive price of only £3,350. A well designed model, the DR 450 fitted the bill for the carriage trade well, although production ceased after a run of only 865 cars, including the two original prototypes and special-bodied cars for heads of state.

At the time of the demise of the Majestic Major Limousine, the only regularly produced limousines on the British market amounted to the staid old Vanden Plas Princess at £3,100 in standard form, produced by the BMC company and itself shortly to go out of production, and the famed Rolls-Royce Phantom V costing an astronomical £10,700 in basic form.

Although other companies would produce the odd stretched version of standard saloons, there was an obvious need for a medium-priced luxury limousine, and both BMC and Jaguar had been considering prospective new models to fill the void; that is until July 1966 when the merger took place between Jaguar Cars Limited and British Motor Holdings Limited. It was obvious to both parties that the market wasn't big enough to support two new limousines, and with the advantage of the group owning the Vanden Plas Company (a renowned name in coachbuilding since 1923) it seemed sensible to combine forces to produce one car and fitting that it should carry the Daimler badge for three reasons: firstly the reputation and prestige of the marque was without question; secondly, because of the relatively low volume production anticipated, Jaguar (on behalf of Daimler) would be best able to build the car to

One of the few limousines available in the 1960s was the Daimler Majestic Major DR 450. With its demise along with the Princess, British Leyland needed a replacement.

141

The original DS 420 limousine design here seen at its launch. With 420G lighting, bumpers, floor pan, boot and much trim it was an economic proposition to produce.

DS 420 production underway at Browns Lane after the move from the Vanden Plas works in Kingsbury, London.

the quality and standard needed; lastly, BMC and Vanden Plas had only been working on an updated version of the old Princess with single piece windscreen and twin headlamps, whereas Jaguar could offer something totally new.

To avoid the astronomical cost of design and re-tooling, it was decided that an existing drive-train would need to be used. Whilst it would have been feasible to adopt the existing 4.5-litre V8 Daimler engine, this would have been very costly as the unit would have been produced in very small quantities for use in one model only, as Jaguar had no intentions of retaining the engine for other saloons. It was therefore decided to utilise the well established Jaguar XK engine in twin carburettor form, developing 165bhp at 4,250rpm, with Borg Warner automatic transmission and Jaguar independent rear suspension. It was a natural progression to consider many of the other standard Jaguar mechanical parts like front suspension, braking system, etc., taken from the Jaguar 420G.

Vanden Plas had been given the task of designing the body of the new limousine and again it was too costly to produce a separate chassis specifically for this model. No suitable chassis existed within the current group range, and this led to the adaptation of Jaguar 420G pressings. They decided to use 420G floorpans (with an extra 20in added in the middle to increase the wheelbase), inner wing panels, scuttle, door pillars, rear pan structure, sills and even boot floor.

It was therefore consistent to utilise external styling features from the 420G. 420G lighting and bumpers featured on the front, the general shaping of the front wings was similar, and hub caps, rear lights and bumper, etc., came directly from the Jaguar. Tyres used, incidentally, were Dunlop H70 HR15 radials. A well designed fluted Daimler radiator grille dominated the front and the final design was certainly very regal and decisively Daimler, taking strong styling features from the 1950s Hooper Empress bodyline. Also, to emulate the "carriage" feel of the car, zero-torque door locks were fitted that allowed the doors to close easily and quietly.

The overall external design was striking yet well balanced, and was much better proportioned than the Majestic Major. Coming in at 18ft 10in long, 6ft 4½in wide, 5ft 3in high and with a wheelbase of 11¾in, the new Daimler was well proportioned by any limousine standards. Despite the enormous bulk of the Daimler and its weight of over 42cwt, with the aid of the Jaguar XK power unit the car was able to progress at a more than stately pace, easily achieving 100mph.

Internally, the car's Jaguar parentage showed in the dashboard, instruments, steering wheel, handles and many aspects of trim.

The standard production model featured a fixed-position front bench seat with glass sliding division. In the rear compartment two occasional foldaway seats were provided, and a separate heating system was operated by controls in one of the armrests. The enormous boot was fully trimmed and fitted with a toolbox taken directly from the 420G model.

Finally designated DS420, the Jaguar-based Daimler limousine was also produced in prototype form as a saloon on a shorter 420G wheelbase to carry the Daimler saloon tradition into the 1970s. This idea, however, was quickly discarded as it had already been decided to introduce a Daimler version of the XJ6 saloon later.

The bodies were made at Jaguar in Coventry and delivered in their bare state to the Vanden Plas coachworks in Kingsbury, London, along with completed engines and transmission. At Vanden Plas the bodies were finished off, painted, trimmed, the mechanics assembled, and the cars tested and delivered to their final destination. By this method of production it was possible to produce running vehicles with incomplete bodies for adaptation to special needs including hearses and cabriolets.

Upon introduction in April 1968 the price of the Daimler DS420 limousine was £2,369 for the chassis only, and

Chauffeur's compartment of an early DS 420 showing the many Jaguar items, and rear compartment of a base model DS 420 with wind-up windows.

£4,425 for the standard limousine. Demand was strong right from the start, and at the car's first public appearance at the 1968 Earls Court Motor Show Jaguar received a Silver Medal for coachwork design.

Being Jaguar based the new Daimler was certainly no sluggard, despite its weight and size. Performance figures recorded by *Motor* revealed:

0-50mph 9.2 secs
0-100mph 43.5 secs
Standing quarter mile 19.5 secs
Top speed 110mph

The limousine was well equipped even by Jaguar standards, but certain items like radio and electric windows were extras. The list of extra-cost options included:

Radio for front and rear compartments
Electrically operated aerial
Electrically operated front and/or rear windows
Electrically operated rear quarter-windows
Electrically operated central glass division
Chromium-plated flag mast
Electric clock in the rear compartment
Adjustable reading lamps in the rear compartment
Nylon rug for the rear compartment
Rubber matting for the front compartment
Wing mirrors
Fire extinguishers
Fog/spot lamps
Seat belts for front seat passenger and driver
Electrically heated rear window
Laminated windscreen
Rear compartment footrests
Chromium-plated badge bar
Sony colour TV in veneered cabinet.

Standard exterior colour schemes were limited to Black and Carlton Grey, but many early cars were also produced in maroon, various shades of grey, navy

The DS 420 was well balanced in style, with the sloping rear boot reminiscent of the 1950s Hooper Empress Line.

blue, greens and whites. Two-tone combinations were not such a rare sight on the DS420, with examples like dark grey over light grey, black over grey, dark blue over light blue and black over maroon.

As well as the above, a host of other options entirely to the owner's taste could be ordered, including:

Fully equipped cocktail cabinet finished in wood veneer
Veneered cabinet for and including colour TV, radio, 4-track stereo
Tinted Sundym glass
Fluorescent lighting in the rear compartment
West of England cloth upholstery for the rear compartment
Various wood veneer finishes
Matching curtains
Reading lamps
Fold-away veneered writing tables
Refrigerated cool box in the boot area
Central division and rear window blinds

A few DS 420s have been supplied with specialised bodies including two landaulette adaptations of the limousine. One of the largest customers for the new limousine was the Daimler Hire Company Limited, who supplied vehicles for many state occasions. Our own Queen, many heads of state as well as Embassies, companies and well known personalities have owned Daimler DS420s, making the car quite a regular sight on the roads of this country.

In 1984 Sir John Egan (then Chief Executive of Jaguar Cars) had an executive-equipped limousine with a fax, an on-board computer link-up via modem, a mobile telephone and a word processor. Priced at that time at £49,000, further examples of this model were available to special order, obviously aimed at captains of industry. In 1970, the Queen Mother took delivery of a DS420 finished in the royal colours of black over claret, without the usual chrome waist-strips (an idea later adopted for all cars), and with her own lion mascot.

Since that time many other fleet users have adopted the DS420, including the Regent Hotel in Hong Kong, which owns 22.

Over the years the DS420 has been mechanically updated in line with Jaguar models, and in 1981 production was transferred to Jaguar's own factory at Browns Lane, Coventry, upon the demise of the VDP works in London. Unfortunately, prior to the move to Coventry, sales were slumping, mainly due to poor quality workmanship and unreliability. Jaguar arranged a massive £3,000 reduction in price to clear the backlog of cars unsold at dealers.

At around the same time (1981) a poorer quality carpet was used to reduce costs, and the General Motors GM400 automatic transmission was fitted in place of the Borg Warner Model 12. The following year heat shields were fitted to protect the interior from heat generated by the exhaust system, and new thermoplastic paint finishes were adopted as for other Browns Lane-produced cars. Out went the Jaguar separate starter button in

For the true executive on the move: DS 420 with TV, on-board computer, telephone, fax ... the works!

favour of a steering column mounted ignition/starter key switch. In order to cut costs further, chromium-plated Rimbellishers were dropped in favour of plain silvered wheels, though still retaining the Daimler hub caps.

In 1983, a better air conditioning system was fitted along with the Jaguar Series 3 engine cooling system and a more powerful American alternator; this upgrading came after many complaints of inadequate ventilation and ineffective engine cooling, particularly in hot countries. Revised boot locks were also fitted at this time, reducing the number of keys needed.

1984 saw the fitting of individual 420G style front seats for chauffeur and passenger; long asked for by drivers, they had individual adjustment for rake and reach, and could be finished in material to suit individual requirements. With this change to the front compartment came the resiting of the automatic transmission quadrant on the floor, requiring a special cloth covered plinth. A fly-off (E-type style!) handbrake was now floor mounted in the centre. At the same time Dunlop 205 tyres replaced the previous 235 Michelins and revised bumpers with smaller rubber faced overriders were fitted, along with rectangular horn grilles and new rectangular indicator lenses, the side lights now being included in the outer headlamps as on other Jaguars.

A revised, quieter exhaust system was fitted from 1985 as well as a manual choke system on the engine. The latter arrived because many of the cars were used for short, slow journeys much of the time, which reduced engine life as the electric choke stayed on.

In 1986 better corrosion treatment was adopted by the factory after many complaints of rusting panelwork. Major styling changes amounted to the adoption of Series 3 saloon bumpers (complete with rubber inserts) to front and rear, with revised fixing mountings to suit. To accompany this change a brand new reduced-depth radiator grille was fitted and, for the first time on a DS420, a Daimler badge appeared on the bootlid. The number of keys was further reduced to two.

For 1988 little changed, except for a reduction in the number of air conditioning grilles in the rear compartment. The following year, however, saw a change in the bodyshell allowing the use of a revised steering column which gave the chauffeur an improved driving position. At the same time out went the plain steel silvered wheels (due to increased short-run production costs)

Revised interior showing Jaguar XJ style steering wheel, XJ gauges and floor mounted auto transmission quadrant.

Perhaps the ultimate in rear compartment conveniences: Sharp colour TV, video, cassette player, and cocktail cabinet on top. Note the optional nylon over-rug.

Final phase of the DS 420 limousine with XJ rubber edged bumper bars and the shallower radiator grille.

in favour of the Kent alloys used on the Daimler Double Six saloon, slightly redesigned to allow the fitment of hub caps.

Over the years, although remaining competitive in the carriage trade market, the DS420 has had problems in matching worldwide legislation, as has the Rolls Royce Phantom VI, and this has inevitably limited its availability overseas. It was also inevitable that these truly handbuilt cars would eventually have to be phased out, and with the demise of both the Daimler and the Rolls-Royce the carriage trade has suffered a great loss, the market now left to "stretched" saloons (like the Jaguar/Daimler XJ40-based Majestic) which don't have the style or comfort of a true limousine.

It was very sad to see the passing of the Daimler DS420 limousine, the last scion of the Mark X, flagship of the Jaguar range from 1961 to 1971. The very last DS420 limousine was completed at the end of 1992, finished in Bordeaux Red with light tan interior, and has been retained by Jaguar as part of its Heritage Collection as a mark of respect for this fine limousine.

A total of 4,981 DS420s were built from 1968 to 1992, the peak year for production being 1970, when a total of 489 were made. Of the grand total, 835 were supplied to outside customers as drive-away chassis for the fitment of hearse and other specialist bodies.

DS420 limousines are still in demand today and one specialist who is possibly the largest trader in limousines in the UK, Worksop Limousines, always offers a wide selection of pre-owned examples from all years. The firm, run by the Hopkinson family in Worksop, North Nottinghamshire, can supply DS420s virtually to order and carry out refurbishment work to the purchaser's specification. They find the Daimlers last well, are economical to run for the carriage trade, reasonably inexpensive secondhand and offer a degree of comfort and prestige not available to their clients at anything like the price elsewhere.

Purchase, Maintenance and Restoration

Gone are the days when the big Jaguars littered the scrapyards... but treasures are still to be found, as in this dusty shed. Such finds can be in remarkably good condition.

Given that readers interested in the models covered in this volume share some common ground, it seems sensible to consider the aspects of purchase and restoration in one chapter. All the cars covered used basically the same basic Jaguar XK power unit and in most cases the same or similar transmissions.

Purchase

With a total of over 77,000 cars built (all versions including DS420s) and despite the ravages of time, there are still plenty of these cars around. Having never been as collectable as the smaller Jaguar saloons or sports cars, prices have remained competitive, making these large, prestigious cars extremely good value for money. The first thing to decide on is which model to go for. Disregarding the DS420 limousines for a moment, the choice must be very much one of personal preference. The Mark VII to IX cars offer the opportunity to enjoy 1950s style motoring, yet with good performance and driveability when compared to many other of the period models. Their poor visibility, archaic driving position and lack of modern creature comforts seem to be the only drawbacks.

The Mark X and 420G on the other hand are perhaps a little more practical, with improved steering and better handling. On the negative side, their bulbous sides and enormous dimensions don't exactly make them ideal town cars and you will need a particu-

148

There was a pull-out drinks tray under the dash.

Engine bay of the 420G saloon. Note the ribbed cam covers and radiator header tank.

Another giant step forward for Jaguar, from Mark X to XJ, in this case a V12 model. The family resemblance is unmistakeable.

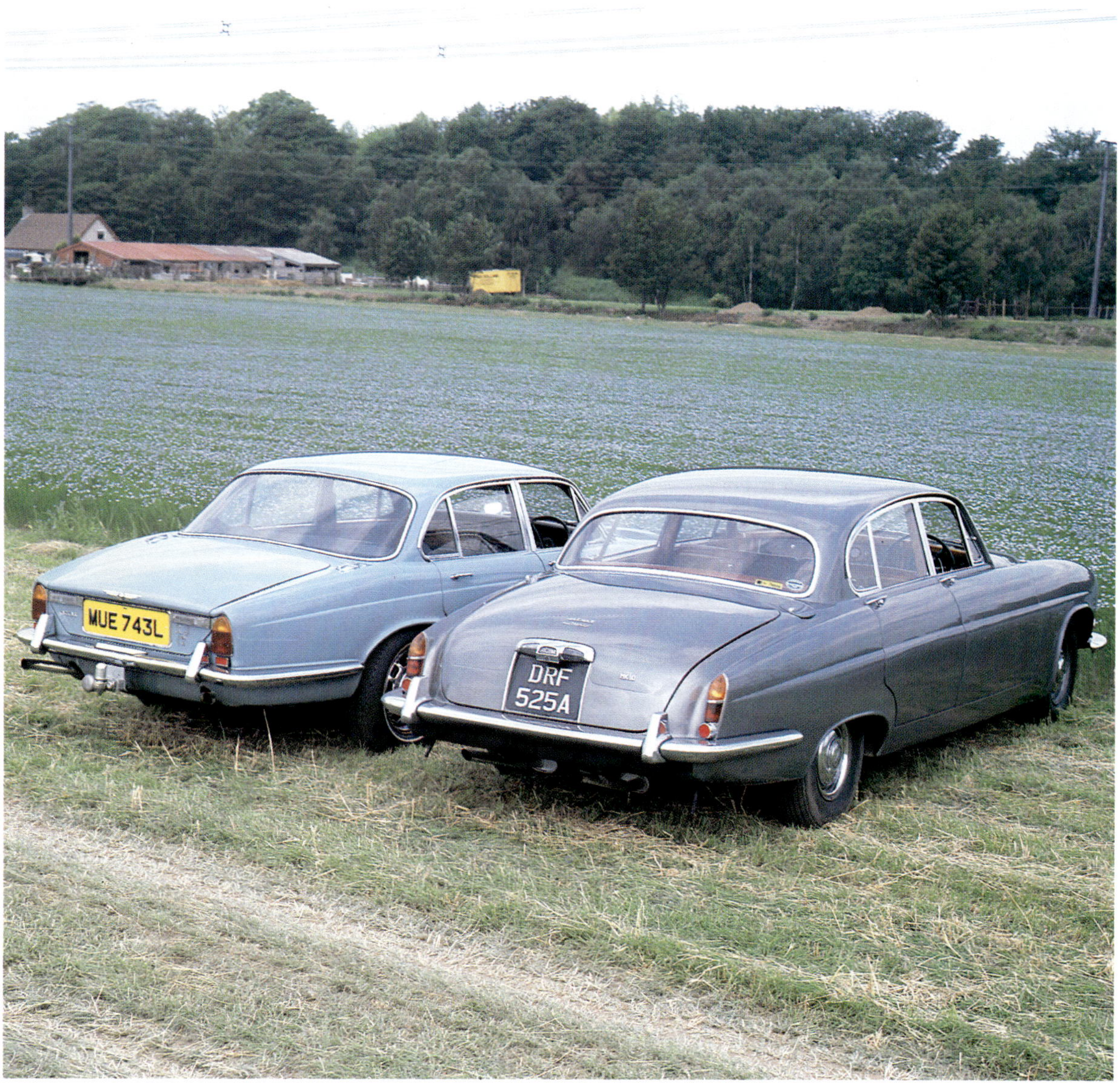
MUE 743L
DRF
525A

The 420G radiator badge and mascot also used on the Mark X.

larly large garage to accommodate one of them.

On price considerations there is little to choose between, say, a concours 420G and a concours Mark IX, although for most cars in average condition the earlier models will inevitably fetch a higher price and are arguably more desirable. When new, the Mark X 420G was never as popular as the earlier Mark VII/IX design and to some extent the same situation applies today. So, on the basis of personal preference, do you want 1950s styling and period feel, or 1960s flamboyance?

If you go for the earlier cars, obviously the Mark IX is the ultimate choice, having the advantages of improved performance from the 3.8-litre power unit, power-assisted steering and all-round disc brakes. A Mark IX can quite happily keep pace with the best of 1960s saloons (and many sports cars) yet has the presence of a true 1950s classic.

The Mark VIII offers the same styling without some of the sophistication but is a distinct rarity due to the small number built. Prices, on the other hand, do not differentiate between the rarity of the Mark VIII and the other models.

If you want purity of design and the original Lyons lines then you must choose the Mark VII. One of the most important Jaguar models, it was built as William Lyons intended, offering cleanness and simplicity of line. The drum brakes are not that bad but the steering will be hard work and you may find the performance disappointing, but there is something about owning the "original" with split screen and bucket seats. The Mark VIIM offers that same purity of style but with improved accommodation and performance.

If the 1960s flamboyance of the Mark X is more to your liking, then here again the later 420G models have proved the most appealing despite the fact that they were only really facelifted versions of the 4.2-litre Mark X. One thing to watch is that some 420Gs were fitted with a 3.77:1 rear axle ratio, which makes them accelerate better but also makes them noisier and thirstier as they run at higher revs. The strength of these cars has meant that they have stood the test of time well, so one is more likely to find a Mark X bodied Jaguar in reasonable condition than most other 1960s luxury cars.

Apart from concours cars, Mark Xs/420Gs can still be found at exceptionally good prices, and there are a surprising number in basically sound and original condition that have not yet been restored. In most cases you are more likely to find the car in urgent need of mechanical work than drastic body surgery. There is little to choose between the Mark X and 420G models but many would consider the early 3.8-litre engined cars the least desirable. They lack the mid-range torque of the later engine and the brakes can be a little suspect, having the old-fashioned Kelsey-Hayes bellows servo.

Whether choosing a Mark VII to IX or a Mark X to 420G, manual transmission versions (preferably with overdrive) are definitely the most desirable, and if you get the added bonus of such items as the period wind-up radio aerials, electric windows (on the Mark Xs only) or even a limousine model, all the better.

As a final alternative there is the Daimler DS420 limousine – although not the ultimate driving experience, it certainly stands out from the crowd. Coachbuilt with style and solid Jaguar engineering, the DS420 can be a surprisingly enjoyable car. Offering a good turn of speed and practicality of maintenance, it is obviously exceptionally roomy and comfortable, an ideal large family vehicle for weekend use and holidays, and you can always put them to work for weddings and local chauffeur work (if you get the appropriate insurance cover).

Being specifically built for the carriage trade DS420s tend to have been looked after reasonably well, although even here corners are sometimes cut to keep the cars on the road as economically as possible, except perhaps in the case of privately owned chauffeur-driven examples. Most will never have been on a motorway, so providing they have been properly and regularly serviced they should be good mechanically! As mentioned earlier, the most likely problems will be with the engine having continually run cold at low speeds. Servicing in such cases should ideally have been carried out every 1,000 miles.

Whilst on the face of it DS420s may look very well kept, the ravages of rust will no doubt have attacked the underside.

Maintenance, restoration and spares availability

Given that the earliest examples of the models covered in this publication are now over 40 years old and even the latest over 20 (except for limousines) the chances of finding one in pristine original condition are small. However, particularly in the case of Mark X bodied cars, it is still possible to find a low-mileage example in good condition if you look hard enough.

Mechanicals

Mechanically, all the cars share the Jaguar XK power unit, a very robust engine which should easily be capable of mileages of up to 150,000 if properly maintained. All XK engines will use or lose oil to a certain extent, depending on mileage and maintenance, but don't be alarmed at figures as low as 200 miles per pint particularly for a 3.8-litre version. Minor loss of oil from the engine is of little concern but major leaks from the rear main oil seal could prove expensive as the engine has to be removed to rectify the fault.

A good oil pressure gauge reading should be a minimum of 40psi when hot at 3,000rpm, and no lower than 15-20psi on tickover. At low revs many higher mileage engines will run at lower pressures but this may not necessarily be detrimental unless the pressure does not rise with engine revs. Engines started from cold should register significantly higher pressures, which should reduce as the oil warms up. Poor oil pressure can be attributable to bearing or oil pump wear, but suspect readings can be put down to the oil pressure relief valve needing cleaning or replacement, or a faulty sender unit.

XK engines should always run reasonably quietly although some tappet noise is to be expected. A rattle from the timing chains is a common fault on Jaguar XK engines and normally becomes most apparent when the engine is run up to 1,500-2,000rpm. The source of the rattle is the front of the engine, and if identified to be from the upper part of the unit it signifies the top timing chain, which is adjustable for wear by way of a special tool (now remade by the Jaguar Enthusiasts' Club). If the noise comes from the lower front of the engine, then this signifies the lower timing chain is at fault, which will prove more costly and time consuming as the engine needs to be removed from the car. The chains can stretch and the tensioners block up or break, needing replacement.

Another engine problem is corrosion of the aluminium alloy cylinder head which, unless very far gone, is not easily identifiable without removal of the head from the block. In extreme cases water will seep into the oil, eventually causing major engine failure. It is vital that Jaguar engines always have the correct level of anti-corrosive anti-freeze maintained in the water system to avoid such corrosion.

Worn valve guides and bore wear can also be a problem, normally identifiable by a blue haze from the exhaust pipes on the over-run. With most Jaguar engines a pale grey haze will be noticed from the exhausts under hard acceleration, which is normal.

It should be mentioned at this point that it is worth checking that the engine numbers tie up with the actual car as it has been commonplace for owners to swap engines around. This not only devalues a car but can also affect the overall performance if the wrong type of engine is fitted. In some cases a 3.8-litre engine may have been fitted to a 3.4 litre Mark VII to improve performance, and if the owner is not concerned with originality he or she can gain the advantages of such a change. Other performance options for the Mark VII can include the fitting of larger carburettors or even at a push a triple-carb unit from either a Mark X or E-Type. In the case of Mark X/420G models it is not unknown for the triple carb unit to be removed (for use in a sports car), to be replaced by an engine from a Mark 2 or 420!

The clutches used on Jaguars are robust items and should not cause trouble providing oil leaks from the engine do not penetrate the plates. It is worth noting that when a clutch needs replacing it is necessary to remove the engine. Any work that requires engine removal requires careful consideration. The engine is big and very heavy, so proper equipment and conditions are required before contemplating such work.

The Moss gearbox (without synchromesh on first gear) was used on Mark VII-IXs and 3.8-litre Mark Xs. Whilst very robust, some parts, particularly layshafts, are not that easy to come by and so can prove expensive to repair. The Moss gearbox is renowned for its slow synchromesh, long travel between

Any mechanical restoration of a Mark VII to X model will involve removal of the engine. You'll definitely need a crane.

first and second gears, and a certain amount of whine (although this should not be excessive). The later, all-synchromesh, Jaguar gearbox as used on the 4.2-litre Mark X and 420G is quieter, faster in operation and just as robust as the earlier Moss 'box.

The Borg Warner automatic transmission used on all the cars covered in this volume is generally very reliable although the earlier DG type can pose some spares problems. In all cases it is vital to keep the filter clean and regularly change the oil. If major repairs are required it is advised that an automatic transmission specialist be approached.

Steering, suspension and brakes should cause no problems, providing regular and proper maintenance has been carried out. The earlier Mark VII to IX models in particular are very easy to work on, having a particularly strong chassis and sound engineering. Proper and regular greasing of all the joints is vital as is adjustment of castor and camber angles.

On the later independent rear suspension models particular care should be given to the rear subframe mounting radius arms, and the inboard disc brakes which can pick up oil leaking from the differential. On the early 3.8-litre Mark X the Kelsey-Hayes servo braking system can be troublesome, with some parts very difficult to obtain. Many convert this model to the later conventional servo system, which is more effective in use.

Possible suspension modifications are numerous and include the lowering of the suspension, fitment of ventilated disc brakes, adjustable shock absorbers, etc., all of which need not necessarily detract too much from originality and do make the car more driveable.

Exhaust systems are not so easy to come by, but for the earlier Mark VII-IX, Mark 2 saloon boxes can be converted, and full stainless steel systems are readily available for all models. As DS420 limousine exhausts are readily available direct from Jaguar these can easily be adapted for use on Mark X/420Gs.

General service items should not cause any difficulties and even tyres are now not that hard to find. The 16in tyres used on Mark VII-IXs can still be found, and the original 6.70 × 16 Dunlop RS5s are now remade, although at a price. Prior to this many owners used to fit Range Rover tyres, which although safe and reasonably successful give a harsh and noisy ride. Tyres for Mark X bodyshell cars used to be a problem as they have 14in rims (an American standard size) and in many cases van standard tyres were fitted, not enhancing safety due to their low speed rating. Nowadays batches of the original Dunlop tyres are being remade. As an alternative many owners have fitted XJ Series 1 15in wheels and tyres. The Mark X/420G hub caps will still fit the wheels and the larger 15in Rimbellishers from early XJs can be used if required.

Bodywork
Generally speaking, the bodies of all the models covered in this book are relatively easy to work on.

Mark Xs, being immensely strong, will have survived best, but as they were of monocoque construction it is vital to check the bodywork thoroughly to ensure there are no structural problems. On the Mark VII bodyshell structural problems are rarely so major, but the large slab-sided panels will inevitably show up repairs unless the work has been done to a very high standard.

Starting with the Mark VII shell, rust and rot can appear in many places. At the front, the area around and below the horn grilles corrodes badly, as does the flat metal valance behind the front

PURCHASE, MAINTENANCE AND RESTORATION

bumper. The front wings rot badly along the rear edge, near the A post, and along the top where the side light pod is leaded into place. The first signs of trouble here show with the leading coming through in a uniform manner around the pod.

As water will have run down through the window channels in the doors, rust may well have formed in the bottoms of the doors, causing the skins and bases to rot out. The central B/C post of the bodyshell is also prone to rusting out at the bottom; it can detach itself from the sill area and therefore cause flexing of the roof and rear doors.

The rear D post, which is double skinned, is a very common area of concern on these shells, rotting out on virtually every car due to a build-up of mud and water thrown up from the rear wheels into the arches.

The sills themselves, although not structural, rot out very easily as do the inner sills where they meet the floors. The floors themselves are also prone to rusting, particularly under the rear seat. The rear wheelarches will rot badly, but this is nicely disguised by the wheel spats normally fitted to the models. The spats themselves are also good mud traps and can cause further rusting. The rear valance under the back bumper also rots out. Underneath the body, the outriggers regularly cause problems, as do rear spring mounting points and rear body mountings.

On the upper parts of the bodyshell, two areas cause concern. Firstly, all bodyshells featured a metal sunroof which, by its very design, allows dirt to collect in the sliding channels, making the roof stiff to open and close. Further dirt builds up in the drain tubes that run along the inside of the roof and down through the rear wings, causing them to block. As the water which is retained eats its way through the tubes and into the roof, severe rotting of the wood-

The bottoms of the front wings where they meet the valance are common areas of rust.

The side light pods on Mark VII-IX front wings were individually welded into place, and the first signs of rotting out will be the lead filler coming through in uniform lines.

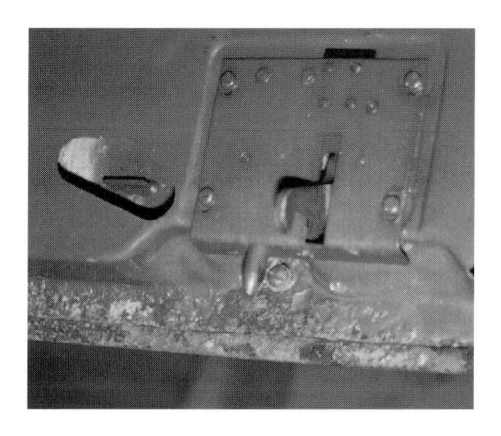

On all models, from VII to X, the bottom of the bootlid is vulnerable to rust...

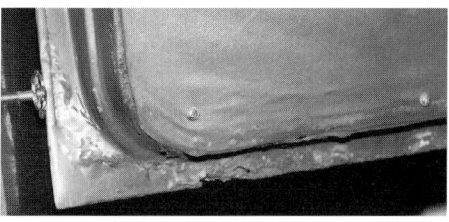

... and so are the door bottoms.

The base of the windscreen on Mark VII-IXs often rusts, as can be seen here.

157

The extreme bottoms of the rear doors on Mark VII-IXs have had plenty of time to rust.

There may be plenty of rot in the double-skinned areas of the rear wheelarches.

Removing the headlining on Mark VII-IX models may reveal unbelievable rust and decay due to condensation.

work frame will take place, eventually showing through the headlining as brown stains. Such problems are prevented by regular maintenance and cleaning of the sunroof mechanisms and channels.

The second area is around the fuel tank filler boxes, one on either side on top of the rear wings. As the rubber seals deteriorate dirt collects, eventually blocking the plastic drain pipes which pass through the rear wings. Water therefore rises into the filler box itself, which rusts away the door hinges to the point where they just fall apart, and it even gets into the petrol tanks. The water drain tubes also block from the bottom, through dirt and mud thrown up off the road. As the tubes are "Y" shaped with the other arm of the "Y" forming the petrol tank breather, water will collect and again find its way into the tanks.

As the inner wings were not painted when new from the factory and the petrol tanks were merely mounted on foam pads, water will collect, rotting away the wings and the tanks. On these fuel tanks a peculiar drain plug system is used, which itself will corrode away giving rise to fuel leaks and smells inside the car.

Hardly any complete replacement panels are available for the Mark VII bodyshells, although side light pods, D post sections, outer and inner sills, rear wheel spats and door bottoms are, but even here, major work is usually needed to get the panels to fit right. At autojumbles the odd "new" Jaguar-made panel can still be found if you look hard enough. Any major remedial work will require the services of a specialist bodybuilder, although it must be stressed that due to the size and design of the bodyshell ease of access should never prove a problem.

Chrome trim is also of major concern on the Mark VII/IX cars. Most items are not available although some gutter and swage line (Mark IX) trims are being remade. Such items as bumpers normally need repairing and replating, as do radiator shells. Remade items common to XK sports cars of the time are available, as well as other items including door handles, mascots, horn grilles and light surrounds. Most electrical items are still readily available, including the previously unobtainable "J" headlights. Such items as hubcaps are common to other contemporary Jaguar models even as late as the 4.2-litre Mark X, but the chromium-plated Rimbellishers are of 16in diameter and not common to many cars of the period. Their fitting clips are also very difficult to come by.

The later monocoque Mark X bodyshells are very strong indeed and despite poor maintenance and hard winters they still remain in relatively good condition. Again virtually no new panels are available, although at autojumbles you can still find prized items. Even fewer parts are available remade than for the earlier Mark VII shell: only

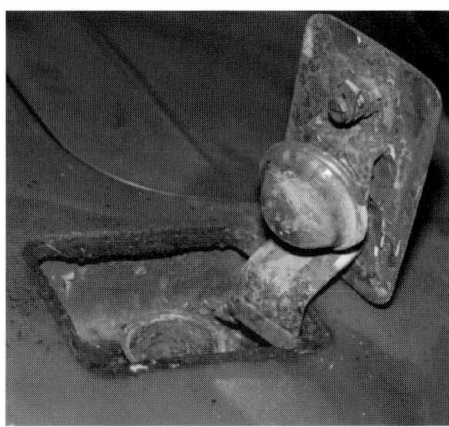

The degree of corrosion around the sunroof area can be extreme.

Fuel filler flaps on all models can conceal a frightful mess.

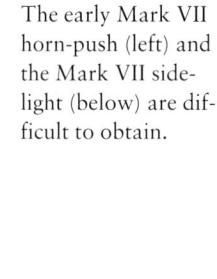

The early Mark VII horn-push (left) and the Mark VII sidelight (below) are difficult to obtain.

Common Mark X problems: rot in rear wings below the bumpers (above), rear wheelarch lips (right) and bottoms of front wings (below right).

inner and outer sills and rear wheel-arch repair sections.

Having said this, the only areas of concern are around the inner and outer wheelarches front and rear, the rear edges of the front wings, the rear wings under the bumpers, sections of the floor panels, the inner and outer sills (structural and made from a thicker gauge steel), the bottoms of all doors, the transverse box sections that stiffen the bodyshell, and inside the bottom of the forward hinged bonnet.

One surprising rust spot on the Mark X bodyshell is the rainwater gutters leading down the windscreen pillars. In severe cases the bottoms of the pillars can rot out completely.

On a more positive note, Mark X doors hardly ever drop despite their immense weight and, unless the car is structurally unsound, there should never be any difficulty in closing the doors.

A lot of Mark X/420G brightwork is common to other Jaguar models; for example the hub caps up to 4.2-litre Mark Xs fit other Jaguar saloons of the period. Later 420G caps were also common to 240/340s, 420s and Series 1 XJs. As with the Mark VII, wheel Rimbellishers are not so easy to find, being of 14ins diameter, and those from other Jaguar saloons of the 1960s (being 15in) will not fit. Boot light nacelles, badges and some lighting equipment are also common. Many other items are now being remade although radiator grilles are a problem as are chrome swage line trims for 420Gs.

DS420s should not be a major concern. Rust in the front wings, rear wheelarches, bottoms of doors and sills is as common on these models as on any other, and new panels can be supplied direct from Jaguar, but at a price – remember that in most respects these were handcrafted limousines. Rear wing repair sections are already available from one specialist company and no doubt other panels will come on the market in due course.

Interior

Consideration has to be given to the interior trim on these cars. There's an awful lot of woodwork and leather and it can be very costly to repair or replace. Over fifty individual pieces of veneer go into a Mark VII and nearly five hides. The veneers wear better than on many later Jaguar saloons and the Vaumol leather upholstery and Wilton carpeting are generally of a higher grade than most. All door trims and non-facing seat trim was of vinyl on all these cars and is thus less costly to replace.

The rear compartment of the Mark VIII and IX models should feature a separate deep-pile nylon over-rug which in many cases will have been damaged or gone missing. It can be costly to have remade but it does complete the picture of luxury.

The wool headlining will have become faded and damaged over the years but is relatively easy to replace, although the flush-fitting sun visors can prove a minor problem. The tool kits fitted in Mark VII to IX front doors sometimes rot out their hinges and the clips that retain the Rexine trim; again these can be difficult to find. The tools themselves can still be found at autojumbles. In most cases they are common to other Jaguar models of the period and some are even being remade.

Instrumentation is in most cases taken from the XK sports cars. Although it is still to be found, it is perhaps better refurbished. On the Mark X/420G cars there are even more pieces of veneered woodwork, and unfortunately the finish does not generally last as long as on the earlier cars. Seat facings also damage more easily, particularly the driver's seat, which seems to sag more than on other models and the

The rusting of fuel filler cover hinges is very common, and finding new replacements (below) is extremely difficult.

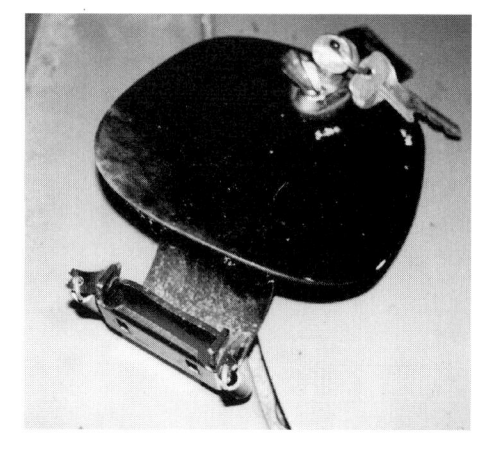

(Below) Mark X doors rarely drop, but one of these has, probably denoting problems in the body structure.

(Bottom) 420G seats had aerated centre panels, expensive to replace.

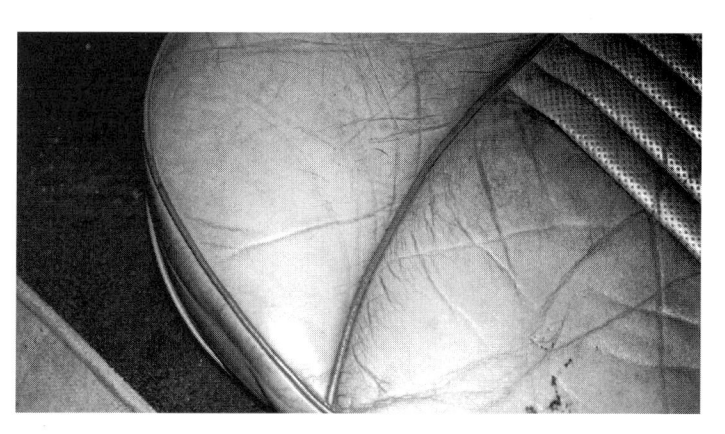

A fine restored or original example of any of the Mark VII to X range of motor cars can be a joy to behold.

top facing of the leather seems to wear away quicker. On 420Gs the leather facing on the seats was of the aerated type, not easily copied if replacement is required. The headlining on later cars was of the sprung board type, which requires a definite knack to fit correctly. Instrumentation is common to all other 1960s Jaguars and is still easy to obtain.

Most models featured Hardura coverings in the boot which are costly to replace. The spare wheel covers for Mark VIII to 420G are particularly prone to discolouration or simply being discarded. The tool kit on the Mark X/420G cars is contained in a metal case in the boot, again with tools common to other Jaguars, although many have unfortunately been raided for parts for the more popular Mark 2s.

It should be noted that within the scope of one chapter it is not possible to cover every single item of concern. Great care should be taken over the purchase of any classic car, and there is no substitute for a thorough investigation of a car with the aid of an expert in the appropriate model. Seek advice and guidance from other owners of similar models and check the intended purchase thoroughly. Never be blinded by a nice shiny paint finish or even copies of bills for work carried out on the car. Join one of the Jaguar (or Daimler) marque clubs to gain further help and guidance, as the enthusiasts you meet may well know of the whereabouts of a car to suit your needs or have experience of someone with whom you intend to deal.

BIG JAGUARS IN COMPETITION

The uninitiated could be forgiven for thinking that Jaguar's entire involvement in motor sport revolves around the Le Mans 24 hour race in the 1950s and a resurgence of competition activity in the 1980s and 90s. Yet Jaguar saloons have seen their fair share of glory, notably the Mark 1s and 2s. Less well known is the success of the Mark VII.

The 1952 Monte Carlo Rally saw the debut of the Mark VII, competing under very bad weather conditions. The Monte itself had been run for a number of years, and an SS 3½-litre saloon made tenth place in 1939. A creditable third place in 1951, with the then obsolete Mark V saloon, followed.

In the 1952 event no less than six Mark VIIs competed, all privately entered except for a factory-loaned car (LWK 343), driven by sports presenter Raymond Baxter with Gordon Wilkins, which unfortunately crashed before the end of the rally. Despite very poor weather conditions the Mark VIIs did well on their first outing, with some excellent results. The best placed Mark VII was fourth, driven by Frenchman René Cotton, followed in sixth position by a Mark VII driven by another Frenchman, Jean Herurtaux. Further down the field at 45th came Bertie Bradnack's Mark VII (severely damaged due to a crash earlier), and then in 53rd place the well-known Yorkshire Jaguar

Norman Dewis test driving the factory Mark VII LWK 343 at MIRA.

Ian Appleyard enthusiastically campaigning his personal Mark VII saloon in the early 1950s.

Stirling Moss at Silverstone for the Daily Express Touring Car Race.

dealer Ian Appleyard in his own Mark VII (PNW 7). Wadham and Waring drove another Mark VII to 15th position, and in consolation won their concours class in the event. Thus, on its first serious competitive outing, the Mark VII behaved impeccably, with no less than five placings, the beginnings of a very successful run.

Mark VIIs competed in other events in 1952, notably, in February, the Daytona Speed Trials in America, when Tom McCahill of *Popular Mechanix* magazine managed to push his Mark

Stirling Moss and the overworked LWK 343 again, coming round for victory at another Silverstone event.

VII to a 100.9mph record despite poor weather and sand conditions.

In the same year Tommy Wisdom drove in the RAC Rally in what was to become the over-worked factory Mark VII, LWK 343. Although he did not get a finishing result he did manage to be fastest in his class. In the Tulip Rally, Ian Appleyard competed in his own Mark VII, taking second place.

The biggest success for the Mark VII in 1952, however, was on home ground at Silverstone, where in the first Production Touring Car Race sponsored by the *Daily Express,* and against competition from the likes of Alvis, Healey, Bristol and even Daimler, Jaguar managed to take first place at an average speed of 75.22mph with LWK 343, driven by none other than Stirling Moss. This particular car, which was to become well known over the next few years, was originally a factory owned standard production car which had been specially prepared by Jaguar's Service Department for this race and for rallies. Using a close-ratio gearbox, tuned engine, 4.27:1 rear axle ratio, XK120 steering and exhaust system and 1in torsion bars, the car was otherwise standard. Not only did it manage first in the 1952 Silverstone event but it also took fastest lap. A further Mark VII driven by Bertie Bradnack finished the event in fourth position.

In 1953, Jaguar attacked the Monte Carlo Rally again, with a record entry of nineteen cars, although unfortunately a Mark VII was not to be the best placed Coventry car, this accolade going to rally veteran Cecil Vard, in his old faithful Mark V saloon, who managed second place overall. The Mark VIIs did perform well, however, with Donald Bennett taking eighth, René Cotton eleventh and Ronnie Adams fifteenth. Ian Appleyard entered again with his wife Pat, but due to unforeseen

Ian Appleyard on the Monte Carlo Rally in his Mark VII.

circumstances only managed to be a runner-up.

In the RAC Rally of the same year, a Mark VII managed to take tenth in class in the hands of Dennis Scott, driving his father's car loaned for the event. In the Tour de France (a particular gruelling rally, requiring a strong vehicle) Novelli and Guido took best in class with their Mark VII. Ian Appleyard entered the Tulip Rally privately and took fifth place with his Mark VII, but crashed in the Norwegian Rally. In the Daily Express Touring Car Race at Silverstone in 1953 Stirling Moss again drove LWK 343, winning for the second year in succession.

One of the Mark VII's more unusual activities in 1953 took place on the Jabbeke Highway in Belgium, where Norman Dewis was already testing the C-Type to achieve a record 148.435mph. At the same time, using LWK 343, Dewis managed to achieve a record 121.704mph for the Mark VII saloon.

The beginning of 1954 saw the Monte Carlo event again, and for the first time Cecil Vard abandoned his Mark V in favour of LWK 343. He finished the rally in eighth position behind Ronnie Adams' Mark VII which took sixth position. Another Mark VII driven by Charles Lampton took the concours award.

In the 1954 RAC Rally J. Ashworth took tenth place and second in class. In the Tulip Rally, Boardman managed a class win and fourth overall. E.R. Parsons performed well in a Mark VII during 1954, with a third in class in the Scottish Rally and an outright win in the Round Britain Rally.

For the Silverstone Touring Car Race later that year Jaguar put up a team of Mark VIIs, with Stirling Moss in LWK 343 and Tony Rolt in another factory car, LHP 5. There were private entries from Ian Appleyard (in his new car registration SUM 7) and Ronnie Adams in his car registered OZ 9499. Very strong competition was experienced from Daimler in the form of their brand new Conquest Century, a smaller, lighter car than the Mark VII. Nevertheless, the Mark VIIs put up a heroic performance, taking the team prize with Appleyard first, Rolt second, Moss third and Ronnie Adams fifth.

For the 1955 Monte Carlo Rally Jaguar gave their official support to three Mark VIIs: Ian Appleyard (SUM 7), Cecil Vard (PWK 701) and Ronnie Adams (PWK 700). Appleyard was working his way through the pack with apparent ease, lying in tenth place at one time, only to drop to 83rd position through water pump problems. Cecil Vard managed 28th and Ronnie Adams took eighth position despite desperate brake fade problems. Jaguar also took the team prize.

Boardman took a class win in the Tulip Rally in 1955, but there were no other major rally activities that year for the Mark VII.

For the 1955 Silverstone Touring Car event Jaguar had assembled a good team of drivers in Ian Appleyard, accompanied by new boys Mike Hawthorn, Jimmy Stewart and Desmond Titterington. Jaguar had already done well in the morning session, racing D-Types, and looked forward to major success with the Mark VIIs after lunch. They were not to be disappointed, with a 1-2-3 win, and a lap record of 2 minutes 10 seconds at 81.6mph achieved by Mike Hawthorn. The final results were: first Mike Hawthorn (LWK 343), second Jimmy Stewart (PWK 700), third Des Titterington (PWK 701). Ian Appleyard had to retire before the finish of the race.

The 1956 Monte Carlo Rally was a momentous occasion for Jaguar as Ronnie Adams took a Mark VII (PWK 700) to outright victory, assisted by drivers Jimmy Stewart, Frank Biggins and Derek Johnson. PWK 700 also took second place in the braking test in the event.

Further success for the Mark VII came at that year's May International Silverstone Meeting, when Ivor Bueb took first place in a Mark VII registered OVC 69, also taking the lap record to 2 minutes 9 seconds. Paul Frère took fourth position in another Mark VII.

With the Suez crisis in 1957 there was little activity on the competition front although a Mark VIII owned by Mrs Anderson, accompanied by Bill Pitt and Jean Abercrombie, won its class in the Round Australia Rally. Another unusual activity for a large Jaguar was the Mobilgas Australian Economy Run, in which a Mark VIII achieved a class win.

With the advent of the successful 2.4- and 3.4-litre Mark 1 saloons, less would be seen of the Mark VIIs as they were by this time considered less competitive. However in 1958, on the RAC rally, Jaguar achieved another 1-2-3 in class with the big saloons, driven by Eric Brinkman, Tommy Sopwith and Tommy Rowe (OVC 69).

No other major successes were to be noted for the Mark VII-bodied cars although private entries continued for some considerable time. Amongst them was Bob Berry, who prepared an ex-factory car (OVC 69) with a lightened mag-

The big Jaguars could still show a clean pair of heels to other classics in saloon car racing in the 1970s and 80s.

The well-known Paladin Jaguar Mark VII, which has competed in many recent events.

Mark VII cornering at Donington in the late 1980s.

nesium alloy bodyshell, modified chassis and D-Type engine. It performed well in the late 1950s and early 60s. The best recorded lap time at Silverstone for a Mark VII in its heyday came from this car, driven by Bob Berry, with a time of 1 minute 19seconds. Later the car was restored by Chris Sturridge and fitted with D-type peg-drive wheels. It competed in many classic car races in the 1970s and is today still in private hands.

Since that time the big Jaguars have seen the race track again in many classic car events, first the Pre-1957 Saloon Car Challenge, then the Classic Saloon Car Challenge and then the Pre-1965 Saloon Car Challenge. Other challenge events included the Jaguar Drivers Club Inter-Area Challenge race series, initially catering for the compact saloons.

Up to 1978 the rules and regulations were very easy, allowing virtually standard production models to compete and giving the club enthusiast the opportunity to propel his Jaguar around a circuit for fun. Since that time, with RAC involvement, the rules have been tightened to improve safety, which has meant that fewer Mark VII-IX cars have been competing. Amongst well known enthusiast drivers Graig Hinton, Henry Crowther, Richard Bradley and Peter Deffee have kept the cars' magic alive. It is interesting to note from the Silverstone Club records that in 1976 Graig Hinton held a lap record in a Mark VIII at 1 minute 18.6 seconds, and again in 1978 in a Mark VII at 1 minute 44.3 seconds. At Brands Hatch similar records were held by C. Nicholson in 1977, and Graig again in 1976 and 1977.

As late as the mid-1980s Mark VII-IXs were still showing a clean pair of heels to the opposition in classic car races. Results in 1983 included:

Classic Saloon Car Challenge: Mark VIII – Henry Crowther first in class.

BIG JAGUARS IN COMPETITION

Pre-'57 Saloon Car Challenge: Mark VIII – Richard Bradley outright winner. Mark VIII – Henry Crowther second in class

Pre-'65 Saloon Car Challenge: Mark VIII – Henry Crowther fourth in class. Mark IX – Dennis Carter sixth in class. Mark VIII – Richard Bradley ninth in class.

Unfortunately, since that time history has repeated itself and the Mark 1 and 2 saloons have proved far more competitive again.

As far as the Mark X bodied cars are concerned, nothing was ever officially seen of the model in its heyday, and even in classic car race circles the car never fared well against the more practical Mark 1s, 2s and S-types. Nevertheless, at least one enthusiastic owner did take up the gauntlet for the Mark X when Roger Wilkinson regularly raced a virtually standard 3.8-litre Mark X (registered 175 BRU) throughout the 1980s. The car was a standard example and Roger merely fitted XJS Kent alloy wheels with low profile tyres, a full harness seat belt and a set of Koni shock absorbers. The car became very competitive and on its first outing at a JDC Silverstone meeting took a creditable sixth fastest overall in practice, managing fifth place in the actual race and second in class. Alas, even Roger Wilkinson in the end abandoned his Mark X in favour of smaller Jaguars, but he nevertheless proved the effectiveness of the Mark X despite its bulk.

Roger Wilkinson's Mark X leaves the grid at Oulton Park in the mid-1980s surrounded by other Jaguars, and in the paddock at Silverstone.

169

Specification, Performance and Production Details

In the following pages the reader will find production figures and technical specifications of the models covered in this book. Performance figures are taken from contemporary road tests (which could vary from car to car and according to weather conditions) or from Jaguar's own information.

MARK VII

Specification

Engine capacity	3,442cc
Number of cylinders	6
Bore and stroke	83 × 106mm
Main bearings	7
Cylinder head	Aluminium alloy, hemispherical combustion chambers, twin overhead camshafts giving 5/16in valve lift, inlet valve diameter 1.75in, exhaust 1.437in
Compression ratio	7:1 or 8:1
Carburettors	Twin 1¾in SU H6
Maximum power	160bhp at 5,200rpm (8:1 cr)
Maximum torque	195lb/ft at 2,500rpm
Steering	Burman recirculating ball
Brakes	Girling hydraulic, 12in drums, servo assisted
Suspension, front	Independent, wishbones, torsion bars, anti-roll bar
Suspension, rear	Live axle with half-elliptic leaf springs
Rear axle	Salisbury hypoid
Overall length	16ft 4½in
Overall width	6ft 1in
Overall height	5ft 3in
Track, front	4ft 8in (4ft 8½in post-1952)
Track, rear	4ft 9½in (4ft 10in post-1952)
Wheelbase	10ft
Turning circle	36ft
Dry weight	33cwt
Tyres and wheels	Dunlop 6.70 × 16 on 16in 5K steel rims (5½k from 1952)

Performance

0-50mph	9.8 secs
0-60mph	13.6 secs
0-90mph	34.4 secs
Standing quarter mile	19.3 secs
Maximum speed	101mph (102mph with overdrive)
Fuel consumption	17.6mpg

Production

12,755 rhd
8,124 lhd
20,929 total
First public announcement October 1955
Produced to end September 1954
First chassis numbers: 710001 rhd, 730001 lhd

MARK VIIM

Specification

Engine capacity	3,442cc
Number of cylinders	6
Bore and stroke	83 × 106mm
Cylinder head	Aluminium alloy, hemispherical combustion chambers, twin overhead camshafts giving 3/8in valve lift, inlet valve diameter 1.75in, exhaust 1.437in. C-Type head also available
Compression ratio	7:1, 8:1 or 9:1
Carburettors	Twin 1¾in SU H6 (2in H8 optional extra)
Maximum power	190bhp at 5,500rpm (8:1 cr)
Maximum torque	203lb/ft at 3,000rpm
Steering	Burman recirculating ball
Brakes	Girling hydraulic, 12in drums, servo assisted
Suspension, front	Independent, wishbones, torsion bars, anti-roll bar
Suspension, rear	Live axle with half-elliptic springs
Rear axle	Salisbury hypoid
Overall length	16ft 4½in
Overall width	6ft 1in
Overall height	5ft 3in
Track, front	4ft 8½in
Track, rear	4ft 10in
Wheelbase	10ft
Turning circle	36ft
Dry weight	33cwt
Tyres and wheels	Dunlop 6.70 × 16 on 5½K steel rims

Performance

0-50mph	9.8 secs
0-60mph	14.1 secs
0-90mph	33.4 secs
Standing quarter mile	19.5 secs

Maximum speed 104.3mph
Fuel consumption 18.8mpg

Production
7,245 rhd
2,016 lhd
9,261 total
First public announcement November 1954
Produced to end July 1957
First chassis numbers: 722755 rhd, 738184 lhd

MARK VIII

Specification
Engine capacity 3,442cc
Number of cylinders 6
Bore and stroke 83 × 106mm
Cylinder head Aluminium alloy, hemispherical combustion chambers, twin overhead camshafts giving ⅜in lift, inlet valve diameter 1.75in, exhaust 1.625in
Compression ratio 7:1, 8:1 or 9:1
Carburettors Twin 1¾in SU HD6
Maximum power 210bhp at 5,500rpm
Maximum torque 216lb/ft at 3,000rpm
Steering Burman recirculating ball
Brakes Girling hydraulic, 12in drums, servo assisted
Suspension, front Independent, wishbones, torsion bars, anti-roll bar
Suspension, rear Live axle, half-elliptic springs
Rear axle Salisbury hypoid
Overall length 16ft 4½in
Overall width 6ft 1in
Overall height 5ft 3in
Track, front 4ft 8½in
Track, rear 4ft 10in
Wheelbase 10ft
Turning circle 36ft
Dry weight 33½cwt
Tyres and wheels Dunlop 6.70 × 16 on 5½K rims

Performance
0-50mph 8.7 secs
0-60mph 11.6 secs
0-90mph 26.7 secs
Standing quarter mile 18.4 secs
Maximum speed 106.5mph
Fuel consumption 17.9mpg

Production
4,644 rhd
1,688 lhd
6,332 total
First public announcement September 1956
Produced concurrently with Mark VIIM at first.
Produced to end December 1959
First chassis numbers: 760001 rhd, 780001 lhd

MARK IX

Specification
Engine capacity 3,781cc
Number of cylinders 6
Bore and stroke 87 × 106mm
Cylinder head Aluminium alloy, hemispherical combustion chambers, twin overhead camshafts giving 0.375in valve lift, inlet valve diameter 1.75in, exhaust 1.625in
Compression ratio 7:1, 8:1 or 9:1
Carburettors Twin SU HD6 1.75in
Maximum power 220bhp at 5,500rpm
Maximum torque 240lb/ft at 3,000rpm
Steering Burman recirculating ball, power assisted
Brakes Dunlop 4-wheel discs, servo assisted
Suspension, front Independent, wishbones, torsion bars, anti-roll bar
Suspension, rear Live axle with half-elliptic springs
Rear axle Salisbury hypoid
Overall length 16ft 4½in
Overall width 6ft 1in
Overall height 5ft 3in
Track, front 4ft 8½in
Track, rear 4ft 10in
Wheelbase 10ft
Turning circle 36ft
Dry weight 35½cwt
Tyres and wheels Dunlop 6.70 × 16 on 5½K steel rims

Performance
0-50mph 8.5 secs
0-60mph 11.3 secs
0-90mph 25.9 secs
0-100mph 34.8 secs
Standing quarter mile 18.1 secs
Maximum speed 114.3mph
Fuel consumption 13.5mpg

Production
5,984 rhd
4,021 lhd
10,005 total
First public announcement October 1958.
Produced concurrently with Mark VII at first.
Produced to end September 1961.
First chassis numbers: 770001 rhd, 79 0001 lhd

MARK X 3.8 LITRE

Specification

Engine capacity	3,781cc
Number of cylinders	6
Bore and stroke	87 × 106mm
Cylinder head	Aluminium alloy, hemispherical combustion chambers, twin overhead camshafts giving ⅜in valve lift, inlet valve diameter 1.75in, exhaust 1.625in
Compression ratio	7:1, 8:1 or 9:1
Carburettors	Triple 2in SU HD8
Maximum power	265bhp at 5,500rpm
Maximum torque	260lb/ft at 4,000rpm
Steering	Burman recirculating ball, power assisted
Brakes	Dunlop 4-wheel discs, inboard at rear, servo assisted
Suspension, front	Independent, wishbones, torsion bars, anti-roll bar
Suspension, rear	Independent, wishbone/upper drivelink, radius arms, coil springs
Rear axle	Salisbury hypoid limited slip
Overall length	16ft 10in
Overall width	6ft 4in
Overall height	4ft 6¾in
Track, front	4ft 8½in
Track, rear	4ft 10in
Wheelbase	10ft
Turning circle	37ft
Dry weight	35cwt
Tyres and wheels	7.50 × 14 Dunlop on 5½K steel rims

Performance

0-50mph	8.4 secs
0-60mph	10.8 secs
0-90mph	24.8 secs
0-100mph	32.9 secs
Standing quarter mile	18.4 secs
Maximum speed	119.5mph
Fuel consumption	13.6mpg

Production

9,129 rhd
3,848 lhd
12,977 total
First public announcement October 1961.
Produced to end August 1964.
First chassis numbers: 300001 rhd, 350001 lhd

MARK X 4.2 LITRE

Specification

Engine capacity	4,235cc
Number of cylinders	6
Bore and stroke	92 × 106mm
Cylinder head	Aluminium alloy, hemispherical combustion chambers, twin overhead camshafts giving ⅜in valve lift, inlet valve diameter 1.75in, exhaust 1.625in
Compression ratio	7:1, 8:1 or 9:1
Carburettors	Triple 2in SU HD8
Maximum power	265bhp at 5,400rpm
Maximum torque	283lb/ft at 4,000rpm
Steering	Burman recirculating ball, with Varamatic power assistance
Brakes	Dunlop 4-wheel discs, inboard at rear, with Dunlop vacuum servo
Suspension, front	Independent, wishbones, torsion bar, anti-roll bar
Suspension, rear	Independent, wishbone/upper drivelink, radius arms, coil springs
Rear axle	Salisbury hypoid, limited slip
Overall length	16ft 10in
Overall width	6ft 4in
Overall height	4ft 6¾in
Track, front	4ft 10in
Track, rear	4ft 10in
Wheelbase	10ft
Turning circle	37ft
Dry weight	36cwt (limousine 37cwt)
Tyres and wheels	7.50 × 14 on 5½K steel rims

Performance

0-50mph	7.9 secs
0-60mph	10.4 secs
0-90mph	22.5 secs
0-100mph	29.5 secs
Standing quarter mile	17.4 secs
Maximum speed	122.5mph
Fuel consumption	16mpg

Production

3,705 rhd saloons
15 rhd limousines
1,957 lhd saloons
3 lhd limousines
5,680 total
First public announcement October 1964.
Produced to end December 1966.
First chassis numbers: ID 50001 rhd, ID 75001 lhd

SPECIFICATION, PERFORMANCE AND PRODUCTION DETAILS

420G

Specification
Engine capacity	4,235cc
Number of cylinders	6
Bore and stroke	92 × 106mm
Cylinder head	Aluminium alloy, hemispherical combustion chambers, twin overhead camshafts giving ⅜in valve lift, inlet valve diameter 1.75in, exhaust 1.625in
Compression ratio	7:1, 8:1 or 9:1
Maximum power	265bhp at 5,400rpm
Maximum torque	283lb/ft at 4,000rpm
Steering	Burman recirculating ball, with Varamatic power assistance
Brakes	Dunlop 4-wheel discs, inboard at rear, with Dunlop vacuum servo
Suspension, front	Independent, wishbones, torsion bars, anti-roll bar
Suspension, rear	Independent, wishbone/ upper drivelink, radius arms, coil springs
Rear axle	Salisbury hypoid, limited slip
Overall length	16ft 10in
Overall width	6ft 4in
Overall height	4ft 6½in
Track, front	4ft 10in
Track, rear	4ft 10in
Wheelbase	10ft
Turning circle	37ft
Dry weight	35¼cwt
Tyres and wheels	Dunlop 205 × 14 radials on 5½K steel rims

Performance
0-50mph	7.2 secs
0-60mph	10.3 secs
0-90mph	22.6 secs
0-100mph	29.9 secs
Standing quarter mile	17 secs
Maximum speed	122.1mph
Fuel consumption	15.8mpg

Production
5,415 rhd saloons
14 rhd limousines
1,115 lhd saloons
10 lhd limousines
6,554 total

DAIMLER DS420

The DS420 is very much a bespoke motor car, so specifications and performance vary. As no road tests have been carried out it must be assumed that performance figures would be slightly worse than those of the Mark X 4.2 litre given the extra weight and greater drag factor.

Specification
Engine capacity	4,235cc
Number of cylinders	6
Bore and stroke	92 × 106mm
Cylinder head	Aluminium alloy, hemispherical combustion chambers, twin overhead camshafts giving ⅜in valve lift, inlet valve diameter 1.75in, exhaust 1.625in
Compression ratio	7:1, 8:1 or 9:1
Carburettors	Twin 2in SU HD8
Maximum power	245bhp at 5,500rpm
Maximum torque	282lb/ft at 3,750rpm
Steering	Burman recirculating ball, with Varamatic power assistance
Brakes	Dunlop 4-wheel discs, inboard at rear, with Girling servo assistance
Suspension, front	Independent, wishbones, torsion bars, anti-roll bar
Suspension, rear	Independent, wishbone/ upper drivelink, radius arms, coil springs
Rear axle	Salisbury hypoid
Overall length	18ft 10in
Overall width	6ft 5½in
Overall height	5ft 3½in
Track, front	4ft 10in
Track, rear	4ft 10in
Wheelbase	11ft 9in
Turning circle	46ft
Dry weight	42cwt
Tyres and wheels	Dunlop H70 HR15 on 5½K steel rims, later Kent alloys

Performance
0-50mph	10.9 secs
Maximum speed	109mph
Fuel consumption	14.8mpg

Production
835 drive-away chassis
4,146 production cars complete with bodies
4,981 total

Useful Addresses

Jaguar Cars Limited
Browns Lane, Allesley, Coventry, Warwickshire
Tel 0203 402121

An active archive department can assist with information on specific cars dependent on proof of current ownership and payment of a small fee. Contact Mrs Ann Harris at Jaguar/Daimler Heritage Trust at the above address for further information.

CLUBS

Jaguar Enthusiasts Club Limited
Graham Searle, Sherborne, Mead Road, Bristol BS12 6TS
Tel 0272 698186

Fastest growing of the Jaguar marque clubs, formed to promote the use of all Jaguar (and Daimler derivative) models. Technical advice, spares and tool remanufacture, discounted insurance and other services.

Jaguar Car Club
Jeff Holman, Barbary, Chobham Road, Horsell, Woking, Surrey GU21 4AS
Tel 0483 763811

Smallest of the Jaguar marque clubs, actively involved in racing, and liaises with other Jaguar marque clubs around the world arranging trips and visits, etc. Also organises two annual Jaguar Spares Days in co-operation with the Jaguar Enthusiasts Club.

Jaguar Drivers Club
18, Stuart Street, Luton, Bedfordshire
Tel 0582 419332

Oldest of the Jaguar marque clubs and pioneered special registers for each Jaguar model.

Daimler and Lanchester Owners Club
Lanchester House, Church Street, Gamlinghay, Sandy, Bedfordshire
Tel 0873 890737

Primarily concerned with the pre-Jaguar Daimlers and Lanchesters, the club now caters for all other Daimler-badged vehicles including the DS420 limousines.

SPARES, REPAIRS, RESTORATIONS, INSURANCE

David Manners
99 Wolverhampton Road, Oldbury, West Midlands B69 4RJ
Tel 021 544 4040 Fax 021 544 5558

Supplier of new and secondhand spares for all models including remanufactured items.

P & K Thornton
Private Road No. 1 Colwick Industrial Estate, Nottingham NG4 2JQ.
Tel 0602 400652 Fax 0602 878011

Restoration including bodywork, lead loading, repaints, repairs, electrical work and re-trimming.

E J Rose
134 Chesterfield Road South, Mansfield, Nottingham NG19 7AP
Tel 0623 24741 Mobile 0860 270909
Fax 0623 640948.

Servicing, repairs and rebuilds of automatic transmissions, overdrives and manual gearboxes.

M & C Wilkinson
Park Farm, Tethering Lane, Everton, nr Doncaster, South Yorkshire DN10 1XX.
Tel 0777 818061 Fax 0777 818049

Suppliers of new and secondhand spares for all models.

J R Transmissions
35A Queens Road, Farnborough, Hampshire
Tel 0252 548337/518508

Repairs and rebuilds of automatic transmissions, manual gearboxes, overdrives and rear axles.

Vintage Tyre Supplies
National Motor Museum, Beaulieu, Hampshire
Tel 0590 612261 Fax 0590 612722

Suppliers of Dunlop original tread pattern tyres for all models.

Aldridge Trimming
St Marks Road, Chapel Ash, Wolverhampton WV3 0QH
Tel 0902 710805/710408 Fax 0902 27474

Interior trim and rubber seal kits for all models.

Useful Addresses

DeWit Classic Jaguar Specialist
Wood End, Elm Walk, Farnborough Park, Orpington, Kent
Tel 0689 856943 Fax 0322 527323

Suppliers of parts for all models.

G W Bartlett and Company
Unit 8, Union Park, Triumph Way, Woburn Road Industrial Estate, Kempston, Bedford MK42 7QB.
Tel 0234 843331 Fax 0234 843340

Suppliers of high quality interior trim for all models.

S C Jaguar Components
13 Cobham Way, Gatwick Road, Crawley, West Sussex
Tel 0293 54781/4 Fax 0293 546570

Falcon and Langford high quality stainless steel exhaust systems.

Norman Motors
100 Mill Lane, London NW6
Tel 071 431 0904 Fax 071 794 5034

Suppliers of new and secondhand spares for all models.

F B Components
35/41 Edgeway Road, Marston, Oxford OX3 0UA
Tel 0865 724646 Fax 0865 250065

Suppliers of new components for all models.

Barry Hankinson
Oxlet House, Bishopswood, Walford, Ross-on-Wye, Hereford HR9 5QX
Tel 0989 65789 Fax 0989 67983

New interior trim, headlinings, etc., for all models.

Flowers Interiors
36 Park Walk, Chase Park, Ross-on-Wye, Hereford HR9 5LW
Tel 0989 62616

Interior trim, carpet sets, etc.

CVS
Unit 1, Longcroft Trade Centre, 209 Glasgow Road, Longcroft, Bonnybridge FK4 1QY
Tel 0324 840133/840858

Jaguar/Daimler dismantlers supplying used parts for all models.

Norton Insurance Brokers
115 Hagley Road, Birmingham B16 8LB
Tel 021 455 6644

Classic car insurance with agreed valuations.

Adam Howell
Croft Street, Walsall, West Midlands WS2 8JR
Tel 0922 649992 Fax 0922 24405

Rechroming service, blast cleaning, polishing, trim repairs, zinc plating.

Black Country Jaguars
Dudley, West Midlands
Tel 0384 456551 Fax 0384 456716

Jaguar/Daimler dismantlers supplying used parts for all models.

A & B Leather and Wood Renovations
1 Dampier Road, Coggeshall, Essex CO6 1QZ
Tel 0376 561586

Refurbishment of leather and veneered woodwork trim.

G H Nolan Ltd
1 St Georges Way, London SE15
Tel 071 701 2785/2669 Fax 071 701 2785

New and secondhand spares for all models.

Coventry Auto Components
Unit 4, Portway Close, Torrington Avenue, Coventry CV4 9UY.
Tel 0203 471217 Fax 0203 421123

Suppliers of XK sports car parts, some of which are interchangeable with Mark VII-IX models.

Colin Webb
31 Elms Drive, Garsington, Oxford OX9 9AG
Tel 086 736 8114/8292 Fax 086 736 8254

Suppliers of service items.

Footman James and Co.
Waterfall Lane, Cradley Heath, Warley, West Midlands B64 6PU
Tel 021 561 4196 Fax 021 559 9203

Classic car insurance with agreed valuations, including Jaguar Enthusiasts Club scheme.

Ken Jenkins Jaguar Spares
Unit 1, High Road, Carlton in Lindrick, Nr Worksop, Nottinghamshire
Tel 0909 732219/730754

Suppliers of new and secondhand parts for all models including remanufactured items and service spares. Technical advisers to Jaguar Enthusiasts Club.

Martin Robey Ltd
Pool Road, Camp Hill Industrial Estate, Nuneaton CV10 9AE
Tel 0203 386903 Fax 0203 345302

Remanufactured panels

Olaf P Lund & Son
2/26 Anthony Road, Saltley, Birmingham B8 3AA
Tel 021 327 2602 Fax 021 327 6284

Suppliers of new and secondhand parts for all models.

David Marks Garages
Rancliffe Garage, Loughborough Road, Bunny, Nottingham
Tel 0602 405370/815143 Mobile 0850 334236

Specialist Jaguar servicing and repairs.

Ken Bell
Crooked Timbers, White Hart Lane, Wood Street Village, Guildford, Surrey
Tel 0483 235153

Specialist Jaguar servicing, repairs and valuations. Technical adviser to the Jaguar Enthusiasts Club.

Brian Reid
9 Barrie Road, Hinckley, Leicestershire

Specialist Jaguar servicing and repairs as well as secondhand parts. Technical adviser to the Jaguar Enthusiasts Club on Mark VII-X models.

Chris Coleman Spares
17 Devonshire Mews, Chiswick, London W4
Tel 081 905 9833

New and secondhand parts.

Jaguar Enthusiasts Club Ltd
Thelma Brotton, Stoneycroft, Moor Lane, Birdwell, Barnsley, South Yorkshire
Tel 0226 742829

Suppliers of remanufactured specialist tools and publications on the Jaguar/Daimler marque.

Baileys UK Ltd
107 Mount Pleasant Road, London NW10

Brake and clutch parts.

The Jag Shop
303 Goldhawk Road, London W12 8EZ
Tel 081 748 7824

Spares.

Ian D Martin
Darlington
Tel 0325 286463

Woodwork refurbishment.

Suffolk and Turley
Unit 7, Attleborough Fields Industrial Estate, Garrett Street, Nuneaton, Warwickshire
Tel 0203 381429

Upholstery restoration and replacement.

Worcester Classic Car Spares
Black and White Cottages, Church Lane, Norton, Worcester WR5 2PS
Tel 0905 821231 Fax 0905 821231

Remanufactured Mark VII-IX panels, and spares.

Bradworthy Classic Restorations
Units 1D and 1E, Langdon Road Industrial Estate, Bradworthy, Holsworthy, Devon
Tel 0409 241791